The Best Technology Writing 2009

Steven Johnson, Editor

The Best Technology Writing 2009

YALE UNIVERSITY PRESS
New Haven and London

Published in association with digitalculturebooks/University of Michigan Press.

Set in Granjon type by The Composing Room of Michigan, Inc., Grand Rapids, Michigan.

Library of Congress Cataloging-in-Publication Data

The best technology writing, 2009 / Steven Johnson, editor.
p. cm.
Includes bibliographical references and index.
ISBN 978-0-300-15410-8 (pbk. : alk. paper)
1. Technology—Social aspects. 2. Information society.
I. Johnson, Steven, 1968–
T14.5.B485 2009
600—dc22 2009011976

A catalogue record for this book is available from the British Library.

Contents

Steven Johnson

Introduction

The most striking thing about the best technology writing of 2009 is how little of it focuses on the future. Since at least the dawn of the digital age, futurists have had the most prominent voices in the tech commentary choir. Their refrain has been: *this is where we are today; now imagine what it'll be like ten years on—or fifty.* But most of the writing included in this volume is grounded squarely in the present. The tacit consensus among the essays you're about to read seems to be: *who needs the future? The present is plenty interesting on its own.*

This present-mindedness has changed the *style* of technology writing. The approach taken in these essays, again and again, is phenomenology not prophecy. They look at the effects of new technology as a real-time, immersive experience, not as a preview of coming attractions. They return, insistently, to the question of what it *feels like* to live in this digital world.

For decades now we've been anticipating the day when mainstream culture would become fully digitized, when the culture of bits and links and downloads would shift from the exception to the rule. But at some point in the past few years, we crossed that Rubicon. I suspect the defining moment was when the generation that grew up assuming that screens were designed to be manipulated and reimagined by their audience

—the generation Doug Rushkoff many years ago dubbed "screenagers"—turned eighteen. It was then that we looked around and realized those kids were now young adults, doing all the things that young adults do today: starting companies from their dorm rooms, updating their Twitter status, playing Grand Theft Auto, and helping to get their political heroes elected.

The ubiquity of the digital lifestyle has forced us to write and think about technology in a different way.

Think back, for example, to Stewart Brand's classic 1973 *Rolling Stone* essay on the first video gamers, "SPACEWAR: Fanatic Life and Symbolic Death Among the Computer Bums." When Brand stumbled across those Stanford proto-gamers, battling each other via command line, it was clear to him that he'd just glimpsed the future. Of course, it took a true visionary like Brand to recognize what he'd encountered, and to write about it with such clarity and infectious curiosity—in the process inventing a whole genre of technology writing that could do justice to the encounter. But there is something about that experience that is also by definition short-sighted: any given technology will mean very different things, and have very different effects, when it is restricted to a small slice of the population. Brand's opening line was "Computers are coming to the people." That was prescient enough. But as it turned out, what he saw on those screens actually had very little to do with the gaming culture we see today. SPACEWAR let Brand sense before just about anyone else that information technology would become as mainstream as rock-and-roll or television. But he couldn't have imagined a culture where games like Spore or World of Warcraft—both of which are deftly dissected in this volume—are far more complex, open-ended, and popular than Hollywood blockbusters.

Likewise, hypertext, until mid-1994, was an emerging

technology whose power users were almost all writers of experimental fiction. You could look at those links on the screen, and begin to imagine what might happen if billions of people started clicking on them. But mostly you were guessing. A shocking amount of the early commentary on hypertext—some of it, in all honesty, written by me—focused on the radical effect hypertext would have on storytelling. Once hypertext went mainstream, however, that turned out to be one of the *least* interesting things about it. (Most of us are still reading novels the old-fashioned way, one page after another.) And that's precisely the trouble with writing about a technology when it's still in leading indicator mode. You could look at those hyperlinks on the screen, and if you really concentrated, you might imagine a future where, say, newspaper articles linked to each other. But you could never imagine Wikipedia or YouPorn.

Now we don't have to imagine it at all: the digital future, to paraphrase William Gibson, is so much more evenly distributed among us. We don't have to gaze into a crystal ball; we can just watch ourselves, self-reflecting as we interact with this vast new ecosystem. Some of my favorite passages in this collection have this introspective quality: the mind examining its own strange adaptation to a world that has been transformed by information technology.

Consider this paragraph, from the opening section of Nicholas Carr's "Is Google Making Us Stupid?": "I'm not thinking the way I used to think. I can feel it most strongly when I'm reading. Immersing myself in a book or a lengthy article used to be easy. My mind would get caught up in the narrative or the turns of the argument, and I'd spend hours strolling through long stretches of prose. That's rarely the case anymore. Now my concentration often starts to drift after two or three pages. I get fidgety, lose the thread, begin looking for something else to do. I feel as if I'm always dragging my wayward brain back to the text. The deep reading that used to come naturally has become a struggle."

Carr intends this as a critique, of course, and his observations will no doubt ring true for anyone who spends hours each day in front of a networked computer screen. I feel it myself right now, as I write this essay, with my open Gmail inbox hovering in the background behind the word processor, and a text message buzzing on my phone, and a whole universe of links tempting me. It *is* harder to sit down and focus on a linear argument or narrative for an hour at a time. In a way, our prophecies about the impact of hypertext on storytelling had it half right; but it's not that people now tell stories using branching hypertext links: it's that *we actively miss those links* when we pick up an old-fashioned book.

Carr is right, too, that there is something regrettable about this shift. The kind of deep, immersive understanding that one gets from spending three hundred pages occupying another person's consciousness is undeniably powerful and essential. And no medium rivals the book for that particular kind of thinking. But it should also be said that this kind of thinking has not simply gone away; people still read books and magazines in vast numbers. It may be *harder* to enter the kind of slow, contemplative state that Carr cherishes, but that doesn't mean it's impossible. I think of our present situation as somewhat analogous to the mass migration from the country to the city that started several centuries ago in Europe: the bustle and stimulation and diversity of urban life made it harder to enjoy the slower, organic pleasures of rural living. Still, those pleasures didn't disappear. People continue to cherish them in mass numbers to this day.

And like urban life, the new consciousness of digital culture has many benefits; it may dull certain cognitive skills, but it undoubtedly sharpens others. In his essay, Carr derides the "skimming" habits of online readers. It's an easy target, particularly when pitted against the hallowed activity of reading a four-hundred-page novel. But skimming is an immensely valuable skill. Most of the information we interact with in our lives—online or off—lacks the profundity and complexity of a Great Book. We don't need deep contemplation to assess an interoffice memo or quarterly financial report from a company we're vaguely interested in. If we can process that information quickly and move on to more important things, so much the better.

Even loftier pursuits benefit from well-developed skimming muscles. I think many of us who feel, unlike Carr, that Google has actually made us *smarter* operate in what I call "skim-and-plunge" mode. We skim through pages of search results or hyperlinked articles, getting a sense of the waters, and then, when we find something interesting, we dive in and read in a slower, more engaged mode. Yes, it is probably a bit harder to become immersed in deep contemplation today than it was sitting in a library in 1985. But that kind of rapid-fire skimming and discovery would have been, for all intents and purposes, impossible before the Web came along.

The benefits of this new consciousness go far beyond skimming, of course, especially when you consider that many of the distractions are not tantalizing hyperlinks but other human beings. Here's Andrew Sullivan describing one of the defining aspects of the experience of blogging, in his revealing essay, "Why I Blog":

> Within minutes of my posting something, even in the earliest days, readers responded. E-mail seemed to unleash their inner beast. They were more brutal than any editor, more persnickety than any copy editor, and more emotionally unstable than any colleague.
>
> Again, it's hard to overrate how different this is. . . . Before the blogosphere, reporters and columnists

> were largely shielded from this kind of direct hazing. Yes, letters to the editor would arrive in due course and subscriptions would be canceled. But reporters and columnists tended to operate in a relative sanctuary, answerable mainly to their editors, not readers. For a long time, columns were essentially monologues published to applause, muffled murmurs, silence, or a distant heckle. I'd gotten blowback from pieces before—but in an amorphous, time-delayed, distant way. Now the feedback was instant, personal, and brutal.

No doubt the intensity and immediacy of the feedback has its own disruptive force, making it harder for the blogger to enter the contemplative state that his forebears in the print magazine era might have enjoyed more easily. Sullivan's description could in fact easily be marshaled in defense of Carr's dumbing-down argument—except that where Carr sees chaos and distraction, Sullivan sees a new kind of engagement between the author and the audience. Sullivan would be the first to admit that this new kind of engagement is noisier, more offensive, and often more idiotic than any traditional interaction between author and editor. But there is so much useful signal in that noise that most of us who have sampled it find it hard to imagine going back. After all, the countryside was more polite, too. But it couldn't compete with the city when it came to the sheer volume of intellectual energy and fun.

Like the city, the Web is populated—annoyingly and beautifully—by people. This has always been the case of course, but it wasn't until the last few years that we began building interfaces that allowed that social dimension to flourish. When most people hear about the personal status updates of Twitter and Facebook for the first time, they think: how inane! But as Clive Thompson describes in his article "Brave New World of Digital Intimacy," the cumulative effect of that distant engagement is much more interesting and subtle: "This is the paradox of ambient awareness. Each little update —each individual bit of social information—is insignificant on its own, even supremely mundane. But taken together, over time, the little snippets coalesce into a surprisingly sophisticated portrait of your friends' and family members' lives, like thousands of dots making a pointillist painting. This was never before possible, because in the real world, no friend would bother to call you up and detail the sandwiches she was eating. The ambient information becomes like 'a type of E.S.P.'"

This experience, too, could be easily entered on Nicholas Carr's side of the ledger: all those status updates distracting us from our deep focus. And, in fact, they *should* be entered on that side of the ledger. Twitter is distracting, no doubt about it. But it's distracting for exactly the reasons Thompson describes: because it's so delightfully human. The truth is Facebook and Twitter belong on *both* sides of the ledger. We have new connections to close friends and utter strangers, a highly attuned social E.S.P., and yet our private reveries can now be interrupted by those same friends and strangers to a degree that would amaze the sleaziest of telemarketers.

As Julian Dibbell and danah boyd show in their essays here, the social web also means that our lives can be disturbed by another class of humans: bullies and cranks, people intent on doing nothing but harassing. The bizarre caste of "griefers" that Dibbell describes was also not visible on the SPACEWAR screens. boyd's essay—originally a blog post—suggests one last form of enlightenment that comes with the mainstreaming of a technology. In the early stages of a new medium, we like to speculate about all the ways that it will change us: how it will make us smarter or more stupid, more socially connected or more isolated, more progressive or more reactionary. But those medium-is-the-message transformations often don't live up to the hype. As boyd points out, the public focus in the outrage over Megan Meier's suicide was on the supposedly new dangers of MySpace rather than the age-old social phenomenon of bullying. "Many legislators are clamoring to make laws based on this case and one thing we know is that bad cases make bad case law," boyd writes. "As in this instance, most legislation and public discourse focus on the technology rather than the damage of psychological abuse and misuse of adult power. . . . And for those who think that bullying is mostly online, think again. The majority of teens believe that bullying is far worse in-person or at school than online." Social network sites don't create more bullying. They just make it more visible to grown-ups. But the fundamental social problem has been there all along. This is the challenge of our present-tense technology writing: most of the time, you have to concentrate on all the ways the digital revolution has transformed our lives. But every now and then, you have to emphasize the continuities, the old messages more than the new medium.

Stewart Brand was right: the computers have come to the people, with all kinds of extraordinary consequences we can now see with a clarity that would have amazed Brand, watching the "computer bums" at Stanford in the early seventies. But sometimes, when the future finally arrives, the most surprising thing you discover is that things aren't that different after all.

Julian Dibbell

Mutilated Furries, Flying Phalluses: Put the Blame on Griefers, the Sociopaths of the Virtual World

Jackassery in Second Life.

The Albion Park section of Second Life is generally a quiet place, a haven of whispering fir trees and babbling brooks set aside for those who "need to be alone to think, or want to chat privately." But shortly after 5 pm Eastern time on November 16, an avatar appeared in the 3-D-graphical skies above this online sanctuary and proceeded to unleash a mass of undiluted digital jackassery. The avatar, whom witnesses would describe as an African-American male clad head to toe in gleaming red battle armor, detonated a device that instantly filled the air with 30-foot-wide tumbling blue cubes and gaping cartoon mouths. For several minutes the freakish objects rained down, immobilizing nearby players with code that forced them to either log off or watch their avatars endlessly text-shout Arnold Schwarzenegger's "Get to the choppaaaaaaa!" tagline from *Predator.*

The incident, it turns out, was not an isolated one. The same scene, with minor variations, was unfolding simultaneously throughout the virtual geography of Second Life. Some cubes were adorned on every side with the infamous, soul-searing "goatse" image; others were covered with the grinning face of Bill Cosby proffering a Pudding Pop.

/Wired/

Soon after the attacks began, the governance team at San Francisco–based Linden Lab, the company that runs Second Life, identified the vandals and suspended their accounts. In the popular NorthStar hangout, players located the offending avatars and fired auto-cagers, which wrapped the attackers' heads in big metallic boxes. And at the Gorean city of Rovere —a Second Life island given over to a peculiarly hardcore genre of fantasy role-play gaming—a player named Chixxa Lusch straddled his giant eagle mount and flew up to confront the invaders avatar-to-avatar as they hovered high above his lovingly re-created medieval village, blanketing it with bouncing 10-foot high Super Mario figures.

"Give us a break you fucks," typed Chixxa Lusch, and when it became clear that they had no such intention, he added their names to the island's list of banned avatars and watched them disappear.

"Wankers," he added, descending into the mess of Super Marios they'd left behind for him to clear.

Bans and cages and account blocks could only slow the attackers, not stop them. The raiders, constantly creating new accounts, moved from one location to another throughout the night until, by way of a finale, they simultaneously crashed many of the servers that run Second Life. And by that time, there was not the slightest mystery in anyone's minds who these particular wankers were: The Patriotic Nigras had struck again.

The Patriotic Nigras consist of some 150 shadowy individuals who, in the words of their official slogan, have been "ruining your Second Life since 2006." Before that, many of them were doing their best to ruin Habbo Hotel, a Finland-based virtual world for teens inhabited by millions of squat avatars reminiscent of Fisher-Price's Little People toys. That's when the PNs adopted their signature dark-skinned avatar with outsize Afro and Armani suit.

Though real-life details are difficult to come by, it's clear that few, if any, PNs are in fact African-American. But their blackface shenanigans, they say, aren't racist in any heartfelt sense. "Yeah, the thing about the racist thing," says ^ban^, leader of the Patriotic Nigras, "is . . . it's all just a joke." It's only one element, he insists, in an arsenal of PN techniques designed to push users past the brink of moral outrage toward that rare moment—at once humiliating and enlightening—when they find themselves crying over a computer game. Getting that response is what it's all about, the Nigras say.

"We do it for the lulz," ^ban^ says—for laughs. Asked how some people can find their greatest amusement in pissing off others, ^ban^ gives the question a moment's thought: "Most of us," he says finally, with a wry chuckle, "are psychotic."

Pwnage, zerging, phat lewts—online gaming has birthed a rich lexicon. But none, perhaps, deserves our attention as much as the notion of the griefer. Broadly speaking, a griefer is an online version of the spoilsport—someone who takes pleasure in shattering the world of play itself. Not that griefers don't like online games. It's just that what they most enjoy about those games is making other players not enjoy them. They are corpse campers, noob baiters, kill stealers, ninja looters. Their work is complete when the victims log off in a huff.

Griefing, as a term, dates to the late 1990s, when it was used to describe the willfully antisocial behaviors seen in early massively multiplayer games like *Ultima Online* and first-person shooters like *Counter-Strike* (fragging your own teammates, for instance, or repeatedly killing a player many levels below you). But even before it had a name, grieferlike behavior was familiar in prehistoric text-based virtual worlds like LambdaMOO, where joyriding invaders visited "virtual rape" and similar offenses on the local populace.

While ^ban^ and his pals stand squarely in this tradition, they also stand for something new: the rise of organized griefing, grounded in online message-board communities and thick with in-jokes, code words, taboos, and an increasingly articulate sense of purpose. No longer just an isolated pathology, griefing has developed a full-fledged culture.

This particular culture's roots can be traced to a semimythic place of origin: the members-only message forums of Something Awful, an online humor site dedicated to a brand of scorching irreverence and gross-out wit that, in its eight years of existence, has attracted a fanatical and almost all-male following. Strictly governed by its founder, Rich "Lowtax" Kyanka, the site boasts more than 100,000 registered Goons (as members proudly call themselves) and has spawned a small diaspora of spinoff sites. Most noticeable is the anime fan community 4chan, with its notorious /b/ forum and communities of "/b/tards." Flowing from this vast ecosystem are some of the Web's most infectious memes and catchphrases ("all your base are belong to us" was popularized by Something Awful, for example; 4chan gave us lolcats) and online gaming's most exasperating wiseasses.

Not all the message boards celebrate the griefers in their midst: Kyanka finds griefing lame, as do many Goons and /b/tards. Nor do the griefers themselves all get along. Patriotic Nigras, /b/tards all, look on the somewhat better-behaved Goon community—in particular the W-Hats, a Second Life group open only to registered Something Awful members—as a bunch of uptight sellouts. The W-Hats disavow any affiliation with the "immature" and "uncreative" Nigras other than to ruefully acknowledge them as "sort of our retarded children."

If there's one thing, though, that all these factions seem to agree on, it's the philosophy summed up in a regularly invoked catchphrase: "The Internet is serious business."

Look it up in the Encyclopedia Dramatica (a wikified lexicon of all things /b/) and you'll find it defined as: "a phrase used to remind [the reader] that being mocked on the Internets is, in fact, the end of the world." In short, "the Internet is serious business" means exactly the opposite of what it says. It encodes two truths held as self-evident by Goons and /b/tards alike—that nothing on the Internet is so serious it can't be laughed at, and that nothing is so laughable as people who think otherwise.

To see the philosophy in action, skim the pages of Something Awful or Encyclopedia Dramatica, where it seems every pocket of the Web harbors objects of ridicule. Vampire goths with MySpace pages, white supremacist bloggers, self-diagnosed Asperger's sufferers coming out to share their struggles with the online world—all these and many others have been found guilty of taking themselves seriously and condemned to crude but hilarious derision.

You might think that the realm of online games would be exempt from the scorn of Goons and /b/tards. How seriously can anyone take a game, after all? And yet, if you've ever felt your cheeks flush with anger and humiliation when some 14-year-old Night Elf in virtual leather tights kicks your ass, then you know that games are the place where online seriousness and online ridiculousness converge most intensely. And it's this fact that truly sets the griefer apart from the mere spoilsport. Amid the complex alchemy of seriousness and play that makes online games so uniquely compelling, the griefer is the one player whose fun depends on finding that elusive edge where online levity starts to take on real-life weight—and the fight against serious business has finally made it seem as though griefers' fun might have something like a point.

History has forgotten the name of the Something Awful Goon who first laid eyes on Second Life, but his initial reaction was undoubtedly along the lines of "Bingo."

It was mid-2004, and Goons were already an organized presence in online games, making a name for themselves as formidable players as well as flamboyantly creative griefers. The Goon Squad guilds in games like *Dark Age of Camelot* and *Star Wars: Galaxies* had been active for several years. In *World of Warcraft,* the legendary Goons of the Mal'ganis server had figured out a way to slay the revered nonplayer character that rules their in-game faction—an achievement tantamount to killing your own team mascot.

But Second Life represented a new frontier in trouble-making potential. It was serious business run amok. Here was an entire population of players that insisted Second Life was not a game—and a developer that encouraged them to believe it, facilitating the exchange of in-game Linden dollars for real money and inviting corporations to market virtual versions of their actual products.

And better still, here was a game that had somehow become the Internet's top destination for a specimen of online weirdo the Goons had long ago adopted as their favorite target: the Furries, with their dedication to role-playing the lives —and sex lives—of cuddly anthropomorphic woodland creatures.

Thus began the Second Life Goon tradition of jaw-droppingly offensive theme lands. This has included the re-creation of the burning Twin Towers (tiny falling bodies included) and a truly icky murdered-hooker crime scene (in which a hermaphrodite Furry prostitute lay naked, violated, and disemboweled on a four-poster bed, while an assortment of coded-in options gave the visitor chances for further violation). But the first and perhaps most expertly engineered of these provocations was Tacowood—a parody of the Furry region known as Luskwood. In Tacowood, rainbow-dappled woodlands have been overrun by the bulldozers and chain saws of a genocidal "defurrestation" campaign and populated with the corpses of formerly adorable cartoon animal folk now variously beheaded, mutilated, and nailed to crosses.

As the media hype around Second Life grew, the Goons began to aim at bigger targets. When a virtual campaign headquarters for presidential candidate John Edwards was erected, a parody site and scatological vandalism followed. When SL real estate magnate Anshe Chung announced she had accumulated more than $1 million in virtual assets and got her avatar's picture splashed across the cover of *BusinessWeek,* the stage was set for a Second Life goondom's spotlight moment: the interruption of a CNET interview with Chung by a procession of floating phalluses that danced out of thin air and across the stage.

People laughed at those attacks, but for Prokofy Neva, another well-known Second Life real estate entrepreneur, no amount of humor or creativity can excuse what she sees as "terrorism." Prokofy (Catherine Fitzpatrick in real life, a Manhattan resident, mother of two, and Russian translator and human-rights worker by trade) earns a modest but bankable income renting out her Second Life properties, and griefing attacks aimed at her, she says, have rattled some tenants enough to make them cancel their leases. Which is why her response to those who defend her griefers as anything but glorified criminals is blunt: "Fuck, this is a denial-of-service attack . . . it's anti-civilization . . . it's wrong . . . it costs me hundreds of US dollars."

Of course, this attitude delights the terrorists in question, and they've made Prokofy a favorite target. The 51-year-old Fitzpatrick's avatar is male, but Goons got ahold of a photo of her, and great sport has been made of it ever since. One build featured a giant Easter Island head of Fitzpatrick spitting out screenshots of her blog. Another time, Prokofy teleported into one of her rental areas and had the "very creepy" experience of seeing her own face looking straight down from a giant airborne image overhead.

Still, even the fiercest of Prokofy's antagonists recognize her central point: Once real money is at stake, "serious business" starts to look a lot like, well, serious business, and messing with it starts to take on buzz-killing legal implications. Pressed as to the legality of their griefing, PNs are quick to cite the distinction made in Second Life's own terms of service between real money and the "fictional currency" that circulates in-game. As ^ban^ puts it, "This is our razor-thin disclaimer which protects us in real-life" from what /b/tards refer to as "a ride in the FBI party van."

Real money isn't always enough to give a griefer pause, however. Sometimes, in fact, it's just a handy way of measuring exactly how serious the griefers' game can get.

Consider the case of the Avatar class Titan, flown by the Band of Brothers Guild in the massively multiplayer deep-space EVE Online. The vessel was far bigger and far deadlier than any other in the game. Kilometers in length and well over a million metric tons unloaded, it had never once been destroyed in combat. Only a handful of player alliances had ever acquired a Titan, and this one, in particular, had cost the players who bankrolled it in-game resources worth more than $10,000.

So, naturally, Commander Sesfan Qu'lah, chief executive of the GoonFleet Corporation and leader of the greater GoonSwarm Alliance—better known outside EVE as Isaiah Houston, senior and medieval-history major at Penn State University—led a Something Awful invasion force to attack and destroy it.

"EVE was designed to be a cold, hard, unforgiving world," explains EVE producer Sígurlina Ingvarsdóttir. It's this attitude that has made EVE uniquely congenial for Goons.

"The ability to inflict that huge amount of actual, real-life damage on someone is amazingly satisfying," says Houston.

"The way that you win in EVE is you basically make life so miserable for someone else that they actually quit the game and don't come back."

And the only way to make someone that miserable is to destroy whatever virtual thing they've sunk the most real time, real money, and, above all, real emotion into. Find the player who's flying the biggest, baddest spaceship and paid for it with the proceeds of hundreds of hours mining asteroids, then blow that spaceship up. "That's his life investment right there," Houston says.

The Goons, on the other hand, fly cheap little frigates into battle, get blown up, go grab another ship, and jump back into the fight. Their motto: "We choke the guns of our enemies with our corpses." Some other players consider the tactic a less-than-sporting end run around a fair fight, still others call it an outright technical exploit, designed to lag the server so the enemy can't move in reinforcements.

Either way, it works, and the success just adds force to GoonFleet's true secret weapon: morale. "EVE is the only game I can think of in which morale is an actual quantifiable source of success," Houston says. "It's impossible to make another person stop playing or quit the game unless their spirit is, you know, crushed." And what makes the Goons' spirit ultimately uncrushable is knowing, in the end, that they're actually playing a different game altogether. As one GoonFleet member's online profile declared, "You may be playing EVE Online, but be warned: We are playing Something Awful."

The Internet is serious business, all right. And of all the ironies inherent in that axiom, perhaps the richest is the fate of the arch-Goon himself, Rich Kyanka. He started Something Awful for laughs in 1999, when he began regularly spotlighting an "Awful Link of the Day." He depends on revenue from SA to sustain not just himself but his pregnant wife, their

2-year-old daughter, two dogs, a cat, and the mortgage on a five-bedroom suburban mini-manor in Missouri. His foothold in the upper middle class rests entirely on the enduring comic appeal of goofy Internet crap.

Sitting in his comfortable basement office at the heart of the Something Awful empire, surrounded by more monitors than the job could possibly require and a growing collection of arch pop-surrealist paintings, Kyanka recounts some of the more memorable moments. Among them: numerous cease-and-desist letters from targets of SA's ridicule, threats of impending bodily harm from a growing community of rage-aholics permabanned from the SA forums, and actual bodily harm from B-movie director Uwe Boll. A onetime amateur boxer, Boll publicly challenged his online critics to a day of one-on-one real-world fights and then pummeled all who showed up, Kyanka among them.

Given that track record, you might think that a family man and sole breadwinner like Kyanka would be looking into another line of work by now. But he's still at it, proudly. "My whole mindset is, there are terrible things on the Internet: Can I write about them and transform them into something humorous?"

But ultimately, Kyanka's persistence is a testament to just how seriously he refuses to take the Internet seriously. Consider: When comments on the Web site of popular tech blogger Kathy Sierra escalated from anonymous vitriol to anonymous death threats last March, it sparked a story that inspired weeks of soul-searching and calls for uniform standards of behavior among bloggers and their communities. In response, Kyanka wrote a Something Awful column, which began with the question: "Can somebody please explain to me how is this news?"

Kyanka went on to review the long and bloodless history of death threats among Internet commenters, then revealed his own impressive credentials as a target: "I've been getting death threats for years now. I'm the king of online dying," he wrote. "Furries hate me, Juggalos hate me, script kiddies hate me, people banned from our forums hate me, people not banned from our forums hate me, people who hate people banned from our forums hate me . . . everybody hates me."

So far, so flip. But almost as an afterthought, Kyanka appended the text of a death threat sent from a banned ex-Goon, aimed not at him but at his infant daughter: "Collateral damage. Remember those words when I kick in your door, duct tape Lauren Seoul's mouth, fuck her in the ass, and toss her over a bridge."

Next to that text, Kyanka posted a photo of himself holding the smiling little girl. His evident confidence in his own safety, and that of the child in his arms, was strangely moving —in an unnerving sort of way.

Moving, and maybe even illuminating. In the end, no matter what they say, life on the Internet really is a serious business. It matters. But the tricky thing is that it matters above all because it mostly doesn't—because it conjures bits of serious human connection from an oceanic flow of words, pictures, videoclips, and other weightless shadows of what's real. The challenge is sorting out the consequential from the not-so-much. And, if Rich Kyanka's steely equanimity is any example, the antics of the Goons and /b/tards might actually sharpen our ability to make that distinction. To those who think the griefers' handiwork is simply inexcusable: Well, being inexcusable is, after all, the griefers' job. Ours is to figure out that caring too much only gives them more of the one thing they crave: the lulz.

Dana Goodyear

I ♥ Novels

Young women develop a genre for the cellular age.

Mone was depressed. It was the winter of 2006, and she was twenty-one, a onetime beauty-school student and a college dropout. She had recently married, and her husband, whom she had known since childhood, was still in school in Tokyo. Thinking that a change might help, she went to stay with her mother, in the country town where she had grown up. Back in her old bedroom, she nursed her malaise, and for weeks she barely left the house. "I'd light a match and see how long it would burn, if you know what I mean," she says. One day, at the end of March, she pulled out all her old photo albums and diaries, and decided to write a novel about her life. She curled up on her side in bed and began typing on her mobile phone.

Mone started posting her novel straight from her phone to a media-sharing site called Maho i-Land (Magic Island), never looking over what she wrote or contemplating plot. "I had no idea how to do that, and I did not have the energy to think about it," she says. She gave her tale a title, "Eternal Dream," and invented, as a proxy for her adolescent self, a narrator named Saki, who is in her second year of high school and lives in a hazily described provincial town. "Where me and my friends live, in the country, there aren't any universities,"

Mone wrote. "If you ride half an hour or so on the train, there's a small junior college, that's all." Saki has a little brother, Yudai, and a close-knit family, a portrait that Mone painted in short, broad strokes: "Daddy / Mom / Yudai / I love you all so much." Before long, however, Saki, walking home from school, is abducted by three strange men in a white car: "—Clatter, clatter—/ The sound of a door opening. / At that moment . . . /—Thud —/ A really dull blunt sound. / The pain that shoots through my head." The men rape her and leave her by the side of the road, where an older boy from school, Hijiri, discovers her. He offers her his jersey, and love is born.

On Mone's third day of writing, readers started to respond. "Please post the next one," and "I'm interested to see what happens," she remembers them writing. She had been posting about twenty screens a day—roughly ten thousand words—divulging as freely as in her diaries, only this was far more satisfying. "Everyone is suffering over their loves and trying to figure out their lives, but my particular struggle was something I wanted to let other girls know about," she says. "Like, 'Hey, girls, I've been through this, you can make it, get up!' "

Soon, Mone's story took a Sophoclean turn. In a sudsy revelation scene, Saki discovers that she is not her father's child. She follows Hijiri to Tokyo for college, but he breaks up with her abruptly. After taking solace in a romance with a younger student named Yuta, she learns that Hijiri is her half brother:

Saki and Hijiri . . .
Are blood relatives, "siblings" . . . ?
The same blood . . .

Runs through our veins . . .
"Older brother and younger sister attract each other"
I've heard something like that.

Around the tenth day, Mone had an epiphany. "I realized that I couldn't put down just exactly what happened," she says. "It came to me that there needed to be the hills and valleys of a story." Her tale acquired the glaze of fiction—how she *wished* her life had turned out rather than how it did. In a postscript, she writes that, unlike Saki and Hijiri, who eventually get back together, Mone and the real Hijiri went their separate ways, and she ended up with Yuta, who had loved her all along.

By mid-April, Mone had completed her novel, nineteen days after she began. Her husband finished school and was starting a job in finance, and she went back to the city to join him. "I was living a casual, unfocussed life in Tokyo," she says, when she heard from Maho i-Land that a publisher wanted to release her novel as a proper book. In December, 2006, "Eternal Dream" was published, at more than three hundred pages. The book distributor Tohan ranked "Eternal Dream" among the ten best-selling literary hardbacks for the first half of 2007. By the end of that year, cell-phone novels, all of them by authors with cutesy one-name monikers, held four of the top five positions on the literary best-seller list. "The Red Thread," by Mei, which has sold 1.8 million copies, was No. 2. "Love Sky," by Mika, was No. 1, and its sequel third; together they have sold 2.6 million copies.

The cell-phone novel, or *keitai shosetsu,* is the first literary genre to emerge from the cellular age. For a new form, it is remarkably robust. Maho i-Land, which is the largest cell-phone-novel site, carries more than a million titles, most of them by amateurs writing under screen handles, and all available for free. According to the figures provided by the company, the site, which also offers templates for blogs and home pages, is visited three and a half billion times a month.

In the classic iteration, the novels, written by and for

young women, purport to be autobiographical and revolve around true love, or, rather, the obstacles to it that have always stood at the core of romantic fiction: pregnancy, miscarriage, abortion, rape, rivals and triangles, incurable disease. The novels are set in the provinces—the undifferentiated swaths of rice fields, chain stores, and fast-food restaurants that are everywhere Tokyo is not—and the characters tend to be middle and lower middle class. Specifically, they are Yankees, a term with obscure linguistic origins (having something to do with nineteen-fifties America and greaser style) which connotes rebellious truants—the boys on motorcycles, the girls in jersey dresses, with bleached hair and rhinestone-encrusted mobile phones. The stories are like folktales, perhaps not literally true but full of telling ethnographic detail. "I suppose you can say *keitai shosetsu* are a source of data or information —the way they use words, how they speak, how they depict scenes," Kensuke Suzuki, a sociologist, told me. "We need these stories so we can learn how young women in Japan commonly feel."

The medium—unfiltered, unedited—is revolutionary, opening the closed ranks of the literary world to anyone who owns a mobile phone. One novelist I met, a twenty-seven-year-old mother of two who lives in the countryside around Kyoto, told me that she thinks up her stories while affixing labels to beauty products at her factory job, and sometimes writes them down on her cell phone while commuting by train to her other job, at a spa in Osaka. But the stories themselves often evince a conservative viewpoint: women suffer passively, the victims of their emotions and their physiology; true love prevails. "From a feminist perspective, for women and girls to be able to speak about themselves is very important," Satoko Kan, a professor who specializes in contemporary women's literature, said. "As a *method,* it leads to the empowerment of girls. But, in terms of content, I find it quite questionable, because it just reinforces norms that are popular in male-dominated culture."

In a country whose low birth rate is a cause for national alarm, and where Tokyo women in their thirties who have yet to find a mate are known as "loser-dogs," the fantasy of rural life offered by the cell-phone novels, with their tropes of teen pregnancy and young love, has proved irresistible. "Love Sky," by Mika, which has been viewed twelve million times online, and has been adapted for manga, television, and film, is a paradigm of sexual mishap and tragedy lightly borne. Freshman year, the heroine—also Mika—falls in love with a rebel named Hiro, and is raped by a group of men incited by Hiro's ex-girlfriend. Then Mika gets pregnant with Hiro's child, and he breaks up with her. Later, she finds out why: he is terminally ill with lymphoma and had hoped to spare her. In the movie version, which came out last fall and earned thirty-five million dollars at the box office, Mika has tears streaming down her face for the better part of two hours. The moral of the story is not that sex leads to all kinds of pain, and so should be avoided, but that sex leads to all kinds of pain, and pain is at the center of a woman's life.

Assuming a pen name is a rite of passage for a writer in Japan. Basho did it in the seventeenth century; Banana Yoshimoto did it in the nineteen-eighties. Mone chose hers rather arbitrarily: she liked the allusion to the French painter and the fact that the Japanese characters can mean "a hundred sounds." But, like many cell-phone novelists, she takes the disguise a great deal further, and makes of her identity a fictional conceit: the spectral, recessive wallflower author whose impression on the world, for all the confessions contained in her novel, is almost illegibly faint. She has a blog, which states her age as eight, her home as "the heart of a mountain," and her hair style as "a poisonous mushroom." Hobbies: drinking, being lazy, and behaving like a baby. Favorite type of guy: the teacher type. There is no photograph, just a cartoon avatar. "I would never let my image be seen," she told me. "If I'm ever photographed, I only show part of my face, just the profile." Apart from her husband, her immediate family, and a few close friends, no one knows that she's the author of "Eternal Dream." "I don't want to bring unwanted attention on my family," she said. "And it's not just me—there's my husband's family to think of, given the things I'm writing. I don't want to inconvenience anyone. Revealing anything, whether it's fiction or truth, is embarrassing, don't you think?"

Mone's withholding is consistent with the ethos of the Japanese Internet, which is dominated by false names and forged identities. "Net transvestites," the most extreme play-actors are called. Match.com doesn't work well here, because a majority of people won't post photographs, and blogs—a recent study found that there are more of them in Japanese than in any other language—are often pseudonymous. Several years ago on 2Channel, a Japanese bulletin-board site that does not require registration, a user started a thread about his unexpected romance with a woman he met on a train. The story, a ballad of Japanese *otaku* (nerd) culture, became "Train Man" —a book, a movie, manga, a television series, and a play—but the author's identity, now hopelessly confused with anonymous collaborators who took the narrative in their own directions, has still not come to light. He is known only as Nakano Hitori: One of Those People. Roland Kelts, a half-Japanese writer born in the United States and the author of "Japanamerica," sees the Internet as an escape valve for a society that can be oppressive in its expectation of normative, group-minded behavior. "In Japan, conflict is not celebrated—consensus is celebrated," he said. "The Internet lets you speak your mind without upsetting the social apple cart." For confessional writers, it is a safe forum for candid self-expression and a magic cloak that makes it easy to disappear into the crowd. "The cell-phone writers have found a pretty clever strategy, through technology, for being part of the culture—participating in that interdependency—and also having a voice," Kelts said.

As an online phenomenon, the novelists were an overlooked subculture, albeit a substantial one. Crossing into print changed that. "In terms of numbers, the fact that the Web had many millions of people accessing and a great number reading is amazing, but the world didn't know whether to praise that or not," Satovi Yoshida, an executive at a cell-phone technology company, said. "With the awful state of publishing, to sell a hundred thousand copies is a big deal. For a previously unpublished, completely unknown author to sell two million copies—that got everyone's attention." Mostly, the attention was negative. In the fall of 2007, Yumi Toyozaki, a popular critic known for her strident reviews, was invited on the radio for what the show's host called a "critical thrashing of the booming cell-phone novels," and brought in a stack of books from the best-seller list.

"No. 10 is 'Eternal Dream,' by—how do you read this name? Mone? Hundred sound?" Toyozaki said. "These names are often formed with just two characters."

"Is the author Chinese-Japanese?" the host asked.

"I don't know."

"It sounds like a dog's name," the host said. He mentioned that these were literary best-sellers.

"I don't even want to use the word 'literary,'" Toyozaki said. "It should be in Other or Yankee."

"I visit a bookstore two or three times a week, but I never stop in the cell-phone-novel area."

Toyozaki concurred. "Once you stop there, you feel sick," she said.

Some feared that the cell-phone novel augured the end of

Japanese literature. "Everyone in publishing received this as an enormous shock to the system, and wondered, What is happening here?" Mikio Funayama, the editor of *Bungakukai,* a respected monthly literary journal, told me. "The author's name is rarely revealed, the titles are very generic, the depiction of individuals, the locations—it's very comfortable, exceedingly easy to empathize with," he said. "Any high-school girl can imagine that this experience is just two steps from her own. But this kind of empathy is largely different from the emotive response—the life-changing event—that reading a great novel can bring about. One tells you what you already know. Literature has the power to change the way you think." For the January, 2008, issue of *Bungakukai,* Funayama assembled a panel to answer the question "Will the cell-phone novel kill 'the author'?" He took some comfort in the panel's conclusion: the novels aren't literature at all but the offspring of an oral tradition originating with mawkish Edo-period marionette shows and extending to vapid J-pop love ballads. "The Japanese have long been attracted to these turgid narratives," he said. "It's not a question of literature being above it. It's just —it's Pynchon versus Tarantino. Most people have a fair understanding of the difference." Banana Yoshimoto, whose extremely popular novels are said to borrow their dreamy, surreal style from girls' manga, wrote in an e-mail, "Youth have their own kind of suffering, and I think that the cell-phone novels became an outlet for their suffering. If the cell-phone novels act as some consolation, that is fine." She went on, "I personally am not interested in them as novels. I feel that it's a waste of time to read them."

Japanese books read right to left, and the script falls vertically from the top of the page, like spiders dangling from silk. The words are combinations of characters drawn from three sources—*hiragana,* a syllabary thought to have been developed for upper-class women, some twelve hundred years ago; *katakana,* a syllabary used mostly for words of foreign origin; and *kanji,* Chinese characters whose mastery is the measure of literary accomplishment. Until the eighties, when the word processor was introduced, the great majority of Japanese was written longhand. (Japanese typewriters, complicated and unwieldy because of all the *kanji,* were left to specialists.) Even now, personal computers are not widespread: one machine per family is common.

For young Japanese, and especially for girls, cell phones —sophisticated, cheap, and, for the past decade, capable of connecting to the Internet—have filled the gap. A government survey conducted last year concluded that eighty-two per cent of those between the ages of ten and twenty-nine use cell phones, and it is hard to overstate the utter absorption of the populace in the intimate portable worlds that these phones represent. A generation is growing up using their phones to shop, surf, play video games, and watch live TV, on Web sites specially designed for the mobile phone. "It used to be you would get on the train with junior-high-school girls and it would be noisy as hell with all their chatting," Yumiko Sugiura, a journalist who writes about Japanese youth culture, told me. "Now it's very quiet—just the little tapping of thumbs." (With the new iPhone and the advent of short-text delivery services like Twitter, American cellular habits are becoming increasingly Japanese; there are at least two U.S. sites, Quillpill and Textnovel, both in the beta stage, that offer templates for writing and reading fiction on cell phones.)

On a Japanese cell phone, you type the syllables of *hiragana* and *katakana,* and the phone suggests *kanji* from a list of words you use most frequently. Unlike working in longhand, which requires that an author know the complex strokes for several thousand *kanji,* and execute them well, writing on a cell phone lowers the barrier for a would-be novelist. The novels are correspondingly easy to read—most would pose no challenge to a ten-year-old—with short lines, simple words, and a repetitive vocabulary. Much of the writing is *hiragana,* and there is ample blank space to give the eyes a rest. "You're not trying to pack the screen," a cell-phone novelist named Rin told me. (Her name, as it happens, actually was borrowed from a dog: her best friend's Chihuahua.) "You're changing the line in the middle of sentences, so where you cut the sentence is an essential part. If you've got a very quiet scene, you use a lot more of those returns and spaces. When a couple is fighting, you'll cram the words together and make the screen very crowded." Quick and slangy, and filled with emoticons and dialogue, the stories have a tossed-off, spoken feel. Satoko Kan, the literature professor, said, "This is the average, ordinary girl talking to herself, the mumblings of her heart."

The Japanese publishing industry, which shrunk by more than twenty per cent over the past eleven years, has embraced cell-phone books. "Everyone is desperately trying to pursue that lifeboat," one analyst told me. Even established publishers have started hiring professionals to write for the market, distributing stories serially (often for a fee) on their own Web sites before bringing them out in print. In 2007, ninety-eight cell-phone novels were published. Miraculously, books have become cool accessories. "The cell-phone novel is an extreme success story of how social networks are used to build a product and launch it," Yoshida, the technology executive, says. "It's a group effort. Your fans support you and encourage you in the process of creating work—they help build the work. Then they buy the book to reaffirm their relationship to it in the first place." In October, the cover of *Popteen,* a magazine aimed at adolescent girls, featured a teenybopper with rhinestone necklaces and pink lipstick and an electric guitar strapped to her chest, wearing a pin that said, "I'd rather be reading."

Printed, the books announce themselves as untraditional, with horizontal lines that read left to right, as on the phone. "The industry saw that there was a new readership," one publishing executive said. "What happens when these girls get older? Will they ever grow up and start reading literature that is vertical? No one knows. But, in a world where everyone is texting and playing games on the Internet, the fact that these paper books are still valued is a good thing." Other conventions established on the screen are faithfully replicated in print. Often, the ink is colored or gray; black text is thought to be too imposing. "Some publishers removed the returns, but those books don't sell well," a representative of Goma Books said. "You need to keep that flow." Goma, which was founded twenty years ago, has emerged as a leading publisher of cell-phone novels. In April, through its Web site, it began releasing for cell phones Japanese literature on which the copyright has expired, including the work of Ryunosuke Akutagawa, Osamu Dazai, and Soseki Natsume. "Masterpieces in your pocket! Read horizontally!" the site declared. This summer, Goma began to print the books in cell-phone style. Its collection of Akutagawa stories, named for his classic short piece "The Spider's Thread," has horizontal blue-gray text and, for cover art, an image of a slender uniformed schoolgirl, lost in thought.

Despite its associations with the nubile and the rustic, the cell-phone novel was invented not by a teen-age girl pining in the provinces but by a Tokyo man in his mid-thirties. Yoshi, as he called himself, was a tutor at a cram school, and later had an office in Shibuya, the hub of youth culture in the nineties: he had plenty of opportunities to observe the beginning of the love affair between young women and their phones. By 2000, when Yoshi set up a Web site and started posting his novel, "Deep Love," Shibuya had been attracting media attention for several years as a center of *enjo kosai,* a form of prostitution in which schoolgirls trade sex with middle-aged men for money or designer clothes. Yoshi's seventeen-year-old heroine sells her body to pay for a heart operation for her boyfriend, Yoshiyuki, but—shades of O. Henry—the money never reaches him, and she dies of AIDS contracted from a client. Yoshi has said that the idea for the heroine's death came from a young reader who wrote to him that she got AIDS from *enjo kosai.* Self-published as a book, "Deep Love" sold a hundred thousand copies.

"This phrase—'a hundred thousand copies'—was what stopped me," Toshiya Arai, the executive director of Starts Publishing Company, said. At the time, Starts, which was founded as a real-estate company, was producing local shopping magazines and dining guides. "I thought this was unprecedented," Arai said. "I thought this person must be a liar, and I wanted to see him face to face." I met with Arai, a small man with sharp eyes and a mole dead center between his brows, and his colleague Shigeru Matsushima in a conference room at the company's office, near Tokyo Station. Arai said that in the summer of 2002 he visited Yoshi, who printed out for him a stack of e-mails from readers. "Nobody was saying that he was a great writer, or that his grammar was good," Arai recalled. "And yet his young fans were all writing about how his book had affected their lives and moved them." A few months later, Starts published "Deep Love," which was made into manga, a television drama, a film, and, eventually, a series of novels that sold 2.7 million copies. "It's a messed-up tale of love," Arai said. "Even among *keitai shosetsu,* it's a sordid adventure." Yoshi, who has left Tokyo and is living quietly in the countryside, has never revealed his name. According to his manager, "Yoshi personally thinks that background information about authors distracts readers when they are reading books."

Around the time that Yoshi started posting, Maho i-Land, which was founded in 1999, added a template called "Let's Make Novels" to its site. After the introduction of unlimited data-transmission packages for cell phones, in 2003, the number of novel writers and readers increased dramatically, an efflorescence as spontaneous as a grow-your-own-crystal set but no less marvellous. Toshiaki Ito, who worked at the company from 2004 to 2007, told me, "By the time I had joined, there was a culture for novels building up on the site. Inside the company, we understood that we had a lot of great content—we had a pile of jewels—and we discussed among ourselves what to do with this treasure chest we had accumulated."

The first of the Maho i-Land trove to be turned into a book was "What the Angel Gave Me," by Chaco, which Starts brought out in 2005. Last year, Arai said, Starts published twenty cell-phone novels, which accounted for nearly a third of the company's forty-three-million-dollar revenue. Mika's "Love Sky" is Starts's most popular title. When I asked Arai if I could meet Mika, he appeared nonplussed. "She is never photographed, and she does not respond to interviews," he said. "This is her wish, and what can we do but honor it? It's understood that the story is based on her experience." I pressed him for details about her identity. "She's twenty-four, and is a woman," he said at last.

"You're not supposed to say her age!" Matsushima, his colleague, snapped. He turned to me. "If you don't mind, could you just say that she's young?"

It was weeks before Mone agreed to see me. When we met, outside a tea shop at a busy intersection not far from Shibuya, she was wearing red tights and Eskimo boots and a meringue-shaped black knit cap with a pompom. Ito, the former Maho i-Land employee, had acted as our liaison, and he was there as a chaperon. As we walked up the street to a traditional Japanese restaurant for dinner, Mone trotted along on the balls of her feet, like a toddler.

Mone is short, with brown hair, curled lashes, and wide-set, placid eyes. She has a bow-shaped mouth and wayward canines—the right one sometimes pokes out through her closed lips, giving her the evil-sweet look of a Nara painting. She was reserved at first, picking daintily at the sashimi course. When a simmering dish of *motsunabe*—cow intestine, cabbage, and tofu—arrived, she took a picture of it with her phone, which was ornamented with a strawberry and a Teddy bear.

As the night progressed, Mone grew more animated. Her literary celebrity had left her feeling bitter—the novel had occasioned heated family fights—but she was mostly angry at herself. "I regret almost everything I've ever published," she said. "I could have done a lot to cover things up and I didn't. I feel a profound responsibility about that." The label of writer, she said, is unsuitable both to her and to the genre. "If I were some super-famous novelist, I would be running around saying, 'Hey, I'm a novelist.' But I'm not. I'm treated as this lame chick who's written one of those awful cell novels. Do you think I can be proud of that? It really depends on which side the public is going to join. I'm considered a total loser for having done it, and I myself think that, too." Her cheeks were flushed, and her eyes glittered. "People say these horrible things about cell-phone novels, and I'm not sure they're mistaken. They say we're immature and incapable of writing a literate sentence. But I would say, so what? The fact that we're producing at all is important."

"Eternal Dream" sold two hundred thousand copies and by now has been accessed nearly three million times online; a sequel, also posted on Maho i-Land, was published in the summer of 2007, and sold eighty thousand copies. Mone calculates that she has made a little less than two hundred thousand dollars from her writing career. At dinner, I asked her if her life had changed in any way. "Not at all," she said. "You have to understand that at no point did I ever think this would feed me," she said. "I'm just another Japanese girl, no better or worse than any other girl walking down the street."

"The Tale of Genji," considered by many to be the world's first novel, was written a thousand years ago, in the Heian period, by a retainer of Empress Akiko at the Imperial Palace, in present-day Kyoto. The Heian was a time of literary productivity that also saw the composition of "The Pillow Book," Sei Shonagon's exquisitely detailed and refined record of court life, and a wealth of tanka poems. We know "Genji"'s author by the name Murasaki Shikibu—Murasaki, or Purple, being the name she gave her story's heroine, and Shikibu the name of the department (Bureau of Ceremonial) where her father at one time worked. Told episodically, and written mostly in *hiragana,* as women at the time were not supposed to learn *kanji,* it is the story of Genji, the beautiful son of the Emperor by a courtesan, who serially charms, seduces, and jilts women, from his rival's daughter to his stepmother and her young niece Murasaki. "Genji" is the epitome of official high culture—it is to the Japanese what the Odyssey is to the Greeks—but some have noticed certain parallels with Japan's new literary boom. "You have the intimate world of the court, and within that you have unwanted pregnancies, people picking on each other, jealousy," the managing director of a large publisher said. "If you simply translate the court for the school, you have the same jealousies and dramas. The structure of 'The Tale of Genji' is essentially the same as a cell-phone novel."

And so it was, in the spirit of continuity, that the third annual Japan Keitai Novel Award, a contest held by Starts, came to have a "Tale of Genji" theme. In late September, fifteen finalists, selected from a pool of thirty-three hundred and fifty who had submitted novels to the Starts Web site, arrived at a big hotel near Tokyo's Imperial Palace for the presentation ceremony. They formed a jittery line in a hallway on the second floor: Saya , in a ruched gray dress, had written "? My vanished love child ? fourteen-year-old pregnancy~ . . . What I really need to tell." White Fig, a graceful young woman with coiffed hair and a netted shawl around her bare shoulders, was the author of "highschoolgirl.co.jp." A doughy, caramel-skinned high-schooler in a sailor suit, clutching a cell phone adorned with hot-pink charms, stood with her parents. She was Kilala, the author of "I want to meet teacher," summarized on the press release as "She loved a man who was her teacher, but already married. Yet the love grows for this kind educator." Kiki ("I'm His Girl")—orangey hair, tartan-print baby-doll dress, pink patent-leather pumps—stomped around with the prize-pony gait of a runway model, and tried to keep her thigh-high stockings from falling down.

The contestants filed into a large ballroom, with pink chrysanthemum-patterned wall-to-wall carpeting, pink chairs, and a shimmering upside-down-wedding-cake chandelier. Strains of dream-sequence harp music filled the air. Seated near the front was Jakucho Setouchi, an eighty-six-year-old novelist and Buddhist nun, who was acting as an honorary judge. The author of searing autobiographical novels in her youth (before she took her vows, her name was Harumi Setouchi), she is a contributor to *Bungakukai* and in the mid-nineties published a contemporary-Japanese translation of "Genji" that became a best-seller. She turned around in her chair to greet the audience: flowing purple robe, white-and-gold brocade *kesa,* shiny bald head.

A government official in a neat suit stood up, and praised the novelists as modern-day Murasakis for their innovative use of 3G cell phones. "The intent of having developed this broadband is for people to use it to create culture, develop new business models, and integrate the provinces into the nation's cultural production," he said. "It's the thousandth anniversary of 'The Tale of Genji.' There was a flowering of culture at that time, and we have hopes that in our new era in Japan we will have the same kind of cultural influence. The authors here are leaders of this new flowering of activity." An announcer on a loudspeaker introduced the finalists, and each stood up and took a shallow bow. "There's one more author, who does not wish to be seen," the announcer added. "She's in the room but doesn't want to be known."

Kiki won the grand prize. When her name was called, she looked startled, and slowly turned her head to the left and to the right, remaining lumpen in her chair. Finally, she advanced to the stage, pulling up her stockings and combing her fingers through her hair. She accepted a huge bouquet from a popular Ping-Pong champion. At the microphone, she wept. She said that she had written the novel for her boyfriend, to commemorate their love. The award was two million yen (some twenty thousand dollars) and publication by Starts.

After Kiki left the stage, laden with the flowers and a signed Ping-Pong paddle, Setouchi made an announcement. Since May, she said, she had been posting a novel on the Starts Web site, under the pen name Purple—the reference to Murasaki Shikibu likely sailed over her readers' heads. Hers was a simple, though well-crafted, tale of a high-school girl, Yuri, who falls in love with a handsome, damaged boy called Hikaru, which is one of Genji's names. Like Genji, Hikaru has an affair with his father's wife and gets her pregnant. (Instead of emperor, Hikaru's father is a corporate executive.) At first, Setouchi said, she had tried to write on her cell phone, but, finding it too difficult, she reverted to her customary medium—traditional Japanese writing paper and a fountain pen—and sent the manuscript to her publisher to convert.

"I'm an author," Setouchi told the audience. "When you finish a novel, to sell tens of thousands would be a tough thing for us, but I see you selling millions. I must confess that I was a bit jealous in the beginning." Then she offered them a word of advice that was probably redundant. "I'm eighty-six years old now, and I don't usually get surprised by things and I don't get so excited, but as long as you're alive you want to be excited, right? But how do you stay excited about life? Keep secrets."

Kiki and her novel were big news. On the social-networking site Mixi, groups organized for and against, debating the merits of her style. The voice of "I'm His Girl" is jivey and loose, unabashedly frank ("Kids? / Well / Twice I got knocked up / By mistake—/ Like who asked them to get made / I / Don't like rubbers / Yeah / For beer and c-cks / Raw is best / You know"), and seasoned with slang expressions, like *se-fure,* for "sex friend," and *mitaina,* a filler word that is the equivalent of "like, you know." The day after the award, a site offering to "convert your blog into the best cell-phone style in 2008" went up on the Internet. All a user had to do was plug in a URL and push a Send key marked "*mitaina*" and the text would be transformed into a snaky uneven column of short lines, punctuated with random occurrences of the word *mitaina.* A message accompanied the translated blogs: "This text was automatically converted to cell-phone novel—here and there line breaks look strange. So what? *Mitaina.*"

Kiki didn't go to college. In high school, she got F's in Japanese. She's twenty-three now, and lives with her boyfriend in a backwater town in Hokkaido, in northern Japan. She has worked in child care, and recently completed a mail-order course in how to look after the elderly. When I spoke to her after her win, she told me that she had written the book because "I was looking back on a difficult thing I had just come through, and I wanted to get it off my chest." She said, "Putting it into this form cleared my mind." In the novel, Aki, the female protagonist, gives up her free-love life style when she falls for a man named Tomo; then she gets pregnant, loses the baby, loses Tomo, and regains Tomo's love at the story's end. Kiki said her real name was similar to her heroine's. "I thought I would be more engaged in the story if her name was close to my own," she said.

I asked Kiki whether she had read "The Tale of Genji." "The problem is the language is so difficult," she said. "There are so many characters." Then she remembered a book she'd read that was a "super-old one, an ancient one!" She said, "I read it four years ago. Before that, I didn't read books of any kind, but it was very easy to read, very contemporary, very close to my life." She told me that the title was "Deep Love."

Farhad Manjoo

The Death of Planned Obsolescence

Why today's gadgets keep getting better. (At least until the battery dies.)

In 2005, a Southern California start-up named Sonos put out a multiroom digital music system, a gadget that sounds straightforward but was actually ahead of its time. Back then, music had already gone digital, but most digital players were meant to be used on the go, not at home. If the iPod is the modern version of the Walkman, Sonos is the reincarnation of the home stereo. It uses wireless networks to string together small "ZonePlayers," stand-alone devices that pipe stereo-quality sound to different rooms in your house. You control the Sonos through a Wi-Fi remote that sports a big LCD screen and an iPod-like scroll wheel. Together, the system's components add up to something transformative: Sonos frees your songs from tinny computer speakers, bringing music to far-flung corners of your McMansion.

But that was three years ago—an eternity in the gadget world. Last week, Sonos offered its first major hardware overhaul since the product's debut (the company decreased the size and increased the networking capabilities of its ZonePlayers). What's remarkable, though, is that while its hardware has barely changed in three years, the Sonos system has improved tremendously since it went on sale. In 2006, the company issued a software update to every Sonos sold—suddenly, the system could play audiobooks. A few months after that, another update allowed Sonos players to hook into the Rhapsody online music service, which meant that for $13 a month, people could now listen to millions of tracks that they didn't own. Later, Sonos added Napster, Pandora, and Sirius, plus a slew of free Internet radio stations. Last year, the company improved its controller's user interface, adding a function that lets you search your tunes from the device—another feature that every Sonos owner got through a software update.

The Sonos isn't cheap—you'll pay $999 for a basic two-room plan, and each additional room will set you back $350 to $500, depending on your hardware needs (the company describes its customer base as "affluent"). But its high price is tempered by a feature that, until recently, was unheard-of in the consumer electronics market: A Sonos you buy today will get *better* as it ages. Through software updates, people who bought the very first Sonos system enjoy pretty much the same functionality that they'd find on a Sonos made two months ago. The company even extends its special offers to its existing customers—last week, both new and current users got a $200 coupon to purchase music from various online services.

Sonos' approach signals a larger shift in the gadget industry, a business that has long titillated its customers with short-lived thrills—what gadget-lovers derisively call "planned obsolescence." It used to be that a gadget worked the best on the day you bought it; every day afterward, it would fall deeper under the shadow of something newer and more fantastic. But because music players, cell phones, cameras, GPS navigators, video game consoles, and nearly everything else now runs on Internet-updatable software, our gadgets' functions are no longer static. It's still true that a gizmo you buy today will eventually be superseded by something that comes along later.

But just like Meryl Streep, your devices will now dazzle you as they age. They'll gain new functions and become easier to use, giving you fewer reasons to jump to whatever hot new thing is just hitting the market.

To appreciate how amazing this is, imagine if the same rules held sway in the car industry. Five years after you bought it, you could take your beater to the shop, and after a quick patch it'd be blessed with electronic stability control, a more fuel-efficient engine, and a radio that received satellite broadcasts.

That sort of metamorphosis is now routine in the consumer electronics business. When Microsoft released the Zune music player late in 2006, critics panned its poor song-beaming feature—you could send tracks to other Zunes, but the music would self-destruct after three days. A year later, Microsoft released a slate of new Zunes. The players featured a more intuitive user interface, and Microsoft dropped the time limit on beamed songs. But here's the kicker: People who'd bought the original Zune also got the new features. A similar thing happened when Apple revamped its original, lame Apple TV set-top box with a less-lame version a few months later. Overnight, a software update gave old Apple TVs the power to buy movies directly from the couch, a feature that had been left out of the first version.

The decline of planned obsolescence is a special boon for start-up companies that aim to break into the market with an entirely new kind of product. A couple of weeks ago, I raved about the Dash GPS navigator, which uses an Internet connection to produce "crowd-sourced" traffic forecasts along your drive. According to the forums on the company's site, there's a lot about the Dash device that people don't like, in particular that its interface is a bit homely, and its traffic detection fails on some roads. But Dash has made its flexibility a key part of its sales pitch: If you're on the fence about the device—if it lacks certain capabilities that you wish it had—the company points out that you won't miss anything by buying now. Your device will eventually get any new functions that are rolled out in new versions.

Of course, there are some features that you can't get through software updates. Because our gadgets are now much like computers, the specs that matter are the same ones we pay attention to when buying PCs—disk space, processor speed, and networking capabilities. For instance, you can expect all future iPods to carry more disk capacity than the one you own today. In the same way, next-generation video game systems will run on much faster processors than are found in today's consoles, and the cell phones of tomorrow will surely include faster wireless Internet speeds than cell phones of today. And one more thing: Eventually the battery in your current phone or PDA or music player will die, and if your device is made by Apple, replacing the battery will be enough of a pain to prompt you to buy something new.

Still, it's surprising how many features can be added to a device without upgrading its hardware. Last month, Apple released the 3G iPhone, which includes faster Internet access than its predecessor, plus GPS access. People who bought the first iPhone can't get those benefits, but they did get what's arguably the best thing in the new iPhone—a software update that allows the device to run third-party applications.

One of these apps magically turns your iPhone into a remote control for iTunes on your computer. I couldn't help thinking of that app as I played around with the fantastic Sonos unit that the company sent me two weeks ago. I fell for the Sonos instantly—the ability to call up any song in any room of your house is hard not to love. But as I played around with the device, I kept thinking of new features I'd like. I want the Sonos to be able to play NPR's Web streams (which can be paused, unlike the Sonos' Internet radio version of NPR).

I'd like the Sonos to act like a DVR, recording certain radio stations at certain times. Mainly, though, I want to be able to control the Sonos through my iPhone, which is much smaller and lighter than the device's own remote.

In an interview, Phil Abram, the company's COO, wouldn't tell me the specific features the Sonos plans to add to its units. But lots of people are asking for an iPhone interface. If the company wants to make its customers happy, it will build one soon—and when that does happen, people who own today's model won't be left out in the cold.

David Talbot

How Obama *Really* Did It

The social-networking strategy that took an obscure senator to the doors of the White House.

Joe Trippi, Howard Dean's 2004 presidential campaign manager and Internet impresario, describes Super Tuesday II—the March 4 primaries in Texas, Ohio, Vermont, and Rhode Island—as the moment Barack Obama used social technology to decisive effect. The day's largest hoard of delegates would be contested in Texas, where a strong showing would require exceptional discipline and voter-education efforts. In Texas, Democrats vote first at the polls and then, if they choose, again at caucuses after the polls close. The caucuses award one-third of the Democratic delegates.

Hillary Clinton's camp had about 20,000 volunteers at work in Texas. But in an e-mail, Trippi learned that 104,000 Texans had joined Obama's social-networking site, www.my.barackobama.com, known as MyBO. MyBO and the main Obama site had already logged their share of achievements, particularly in helping rake in cash. The month before, the freshman senator from Illinois had set a record in American politics by garnering $55 million in donations in a single month. In Texas, MyBO also gave the Obama team the instant capacity to wage fully networked campaign warfare. After seeing the volunteer numbers, Trippi says, "I remember saying, 'Game, match—it's over.'"

The Obama campaign could get marching orders to the Texans registered with MyBO with minimal effort. The MyBO databases could slice and dice lists of volunteers by geographic micro-region and pair people with appropriate tasks, including prepping nearby voters on caucus procedure. "You could go online and download the names, addresses, and phone numbers of 100 people in your neighborhood to get out and vote—or the 40 people on your block who were undecided," Trippi says. "'Here is the leaflet: print it out and get it to them.' It was you, at your computer, in your house, printing and downloading. They did it all very well." Clinton won the Texas primary vote 51 to 47 percent. But Obama's people, following their MyBO playbook, so overwhelmed the chaotic, crowded caucuses that he scored an overall victory in the Texas delegate count, 99 to 94. His showing nearly canceled out Clinton's win that day in Ohio. Clinton lost her last major opportunity to stop the Obama juggernaut. "In 1992, Carville said, 'It's the economy, stupid,'" Trippi says, recalling the exhortation of Bill Clinton's campaign manager, James Carville. "This year, it was the network, stupid!"

Throughout the political season, the Obama campaign has dominated new media, capitalizing on a confluence of trends. Americans are more able to access media-rich content online; 55 percent have broadband Internet connections at home, double the figure for spring 2004. Social-networking technologies have matured, and more Americans are comfortable with them. Although the 2004 Dean campaign broke ground with its online meeting technologies and blogging, "people didn't quite have the facility," says Lawrence Lessig, a Stanford law professor who has given the Obama campaign Internet policy advice. "The world has now caught up with the technology." The Obama campaign, he adds, recognized this early:

"The key networking advance in the Obama field operation was really deploying community-building tools in a smart way from the very beginning."

Of course, many of the 2008 candidates had websites, click-to-donate tools, and social-networking features—even John McCain, who does not personally use e-mail. But the Obama team put such technologies at the center of its campaign—among other things, recruiting 24-year-old Chris Hughes, cofounder of Facebook, to help develop them. And it managed those tools well. Supporters had considerable discretion to use MyBO to organize on their own; the campaign did not micromanage but struck a balance between top-down control and anarchy. In short, Obama, the former Chicago community organizer, created the ultimate online political machine.

The Obama campaign did not provide access or interviews for this story; it only confirmed some details of our reporting and offered written comments. This story is based on interviews with third parties involved in developing Obama's social-networking strategy or who were familiar with it, and on public records.

AN ONLINE NERVOUS SYSTEM

A row of elegant, renovated 19th-century industrial buildings lines Boston's Congress Street east of Fort Point Channel. On any given day, behind a plain wooden door on the third floor of 374 Congress, 15 to 20 casually clad programmers tap away at computers. On the day I visited, the strains of Creedence Clearwater Revival filled the room; a Ping-Pong table dominated the small kitchen. This is the technology center for Blue State Digital, which means that it is also the nervous system for its two largest clients, the Barack Obama campaign and the Democratic National Committee. Founded by alumni of the Dean campaign, Blue State Digital added interactive elements to Obama's website—including MyBO—and now tends to its daily care and feeding. The site's servers hum away in a Boston suburb and are backed up in the Chicago area.

Jascha Franklin-Hodge, 29, greeted me with a friendly handshake and a gap-toothed grin. He has a deep voice and a hearty laugh; his face is ringed by a narrow beard. Franklin-Hodge dropped out of MIT after his freshman year and spent a few years in online music startups before running the Internet infrastructure for the Dean campaign, which received a then-unprecedented $27 million in online donations. "When the campaign ended, we thought, 'Howard Dean was not destined to be president, but what we are doing online—this is too big to let go away,'" he says. He and three others cofounded Blue State Digital, where he is chief technology officer. (Another cofounder, Joe Rospars, is now on leave with the Obama campaign as its new-media director.)

The MyBO tools are, in essence, rebuilt and consolidated versions of those created for the Dean campaign. Dean's website allowed supporters to donate money, organize meetings, and distribute media, says Zephyr Teachout, who was Dean's Internet director and is now a visiting law professor at Duke University. "We developed all the tools the Obama campaign is using: SMS [text messaging], phone tools, Web capacity," Teachout recalls. "They [Blue State Digital] did a lot of nice work in taking this crude set of unrelated applications and making a complete suite."

Blue State Digital had nine days to add its tools to Obama's site before the senator announced his candidacy on February 10, 2007, in Springfield, IL. Among other preparations, the team braced for heavy traffic. "We made some projections of traffic levels, contribution amounts, and e-mail levels based on estimates from folks who worked with [John] Kerry and Dean in 2004," recalls Franklin-Hodge. As Obama's

Springfield speech progressed, "we were watching the traffic go up and up, surpassing all our previous records." (He would not provide specific numbers.) It was clear that early assumptions were low. "We blew through all of those [estimates] in February," he says. "So we had to do a lot of work to make sure we kept up with the demand his online success had placed on the system." By July 2008, the campaign had raised more than $200 million from more than a million online donors (Obama had raised $340 million from all sources by the end of June), and MyBO had logged more than a million user accounts and facilitated 75,000 local events, according to Blue State Digital.

MyBO and the main campaign site made it easy to give money—the fuel for any campaign, because it pays for advertising and staff. Visitors could use credit cards to make one-time donations or to sign up for recurring monthly contributions. MyBO also made giving money a social event: supporters could set personal targets, run their own fund-raising efforts, and watch personal fund-raising thermometers rise. To bring people to the site in the first place, the campaign sought to make Obama a ubiquitous presence on as many new-media platforms as possible.

The viral Internet offered myriad ways to propagate unfiltered Obama messages. The campaign posted the candidate's speeches and linked to multimedia material generated by supporters. A music video set to an Obama speech—"Yes We Can," by the hip-hop artist Will.i.am—has been posted repeatedly on YouTube, but the top two postings alone have been viewed 10 million times. A single YouTube posting of Obama's March 18 speech on race has been viewed more than four million times. Similarly, the campaign regularly sent out text messages (at Obama rallies, speakers frequently asked attendees to text their contact information to his campaign) and made sure that Obama was prominent on other social-networking sites, such as Facebook and MySpace. The campaign even used the micro-blogging service Twitter, garnering about 50,000 Obama "followers" who track his short posts. "The campaign, consciously or unconsciously, became much more of a media operation than simply a presidential campaign, because they recognized that by putting their message out onto these various platforms, their supporters would spread it for them," says Andrew Rasiej, founder of the Personal Democracy Forum, a website covering the intersection of politics and technology (and another Dean alumnus). "We are going from the era of the sound bite to the sound blast."

Money flowed in, augmenting the haul from big-ticket fund-raisers. By the time of the Iowa caucuses on January 3, 2008, the Obama campaign had more than $35 million on hand and was able to use MyBO to organize and instruct caucus-goers. "They have done a great job in being precise in the use of the tools," Teachout says. "In Iowa it was house parties, looking for a highly committed local network. In South Carolina, it was a massive get-out-the-vote effort." MyBO was critical both in the early caucus states, where campaign staff was in place, and in later-voting states like Texas, Colorado, and Wisconsin, where "we provided the tools, remote training, and opportunity for supporters to build the campaign on their own," the Obama campaign told *Technology Review* in a written statement. "When the campaign eventually did deploy staff to these states, they supplemented an already-built infrastructure and volunteer network."

Using the Web, the Obama camp turbocharged age-old campaign tools. Take phone banks: through MyBO, the campaign chopped up the task of making calls into thousands of chunks small enough for a supporter to handle in an hour or two. "Millions of phone calls were made to early primary states by people who used the website to reach out and connect with them," Franklin-Hodge says. "On every metric, this campaign has operated on a scale that has exceeded what has been done before. We facilitate actions of every sort: sending e-mails out to millions and millions of people, organizing tens of thousands of events." The key, he says, is tightly integrating online activity with tasks people can perform in the real world. "Yes, there are blogs and Listservs," Franklin-Hodge says. "But the point of the campaign is to get someone to donate money, make calls, write letters, organize a house party. The core of the software is having those links to taking action—to doing something."

PORK INVADERS

If the other major candidates had many of the same Web tools, their experiences show that having them isn't enough: you must make them central to the campaign and properly manage the networks of supporters they help organize. Observers say that Clinton's campaign deployed good tools but that online social networks and new media weren't as big a part of its strategy; at least in its early months, it relied more on conventional tactics like big fund-raisers. After all, Clinton was at the top of the party establishment. "They [the Obama supporters] are chanting 'Yes we can,' and she's saying 'I don't need you,'" Trippi says. "That is what the top of that campaign said by celebrating Terry McAuliffe [the veteran political operative and former Democratic National Committee chairman] and how many millions he could put together with big, big checks. She doesn't need my $25!" The two campaigns' fund-raising statistics support Trippi's argument: 48 percent of Obama's funds came from donations of less than $200, compared with 33 percent of Clinton's, according to the Center for Responsive Politics.

Clinton's Internet director, Peter Daou, credits the Obama campaign with doing an "amazing job" with its online social network. "If there is a difference in how the two campaigns approached [a Web strategy], a lot of those differences were based on our constituencies," Daou says. "We were reaching a different demographic of supporters and used our tools accordingly." For example, he says, the Clinton campaign established a presence on the baby-boomer social-networking site Eons.com, and Clinton herself often urged listeners to visit www.hillaryclinton.com. But Andrew Rasiej says that the conventional political wisdom questioned the value of the Internet. "As far as major political circles were concerned," he says, "Howard Dean failed, and therefore the Internet didn't work."

While it's hard to tease out how much Clinton's loss was due to her Web strategy—and how much to factors such as her Iraq War vote and the half-generation difference between her and Obama's ages—it seems clear that her campaign deëmphasized Web strategy early on, Trippi says. Even if you "have all the smartest bottom-up, tech-savvy people working for you," he says, "if the candidate and the top of the campaign want to run a top-down campaign, there is nothing you can do. It will sit there and nothing will happen. That's kind of what happened with the Clinton campaign."

Republican Ron Paul had a different problem: Internet anarchy. Where the Obama campaign built one central network and managed it effectively, the Paul campaign decided early on that it would essentially be a hub for whatever networks the organizers were setting up. The results were mixed. On the one hand, volunteers organized successful "money bombs"—one-day online fund-raising frenzies (the one on November 5, 2007, netted Paul $4.3 million). But sometimes the volunteers' energy—and money—was wasted, says Justine Lam, the Paul campaign's Internet director, who is now the online marketing director at Politicker.com. Consider the supporter-driven effort to hire a blimp emblazoned with "Who is Ron Paul? Google Ron Paul" to cruise up and down the East Coast last winter. "We saw all this money funding a blimp, and thought, 'We really need this money for commercials,'" Lam says.

Then there is McCain, who—somewhat ironically—was the big Internet story of 2000. That year, after his New Hampshire primary victory over George W. Bush, he quickly raised $1 million online. And at times last year, he made effective use of the Internet. His staff made videos—such as "Man in the Arena," celebrating his wartime service—that gained popularity on YouTube. But the McCain site is ineffectual for social networking. In late June, when I tried to sign up on McCainSpace—the analogue to MyBO—I got error messages. When I tried again, I was informed that I would soon get a new password in my in-box. It never arrived. "His social-networking site was poorly done, and people found there was nothing to do on it," says Lam. "It was very insular, a walled garden. You don't want to keep people inside your walled garden; you want them to spread the message to new people."

McCain's organization is playing to an older base of supporters. But it seems not to have grasped the breadth of recent shifts in communications technology, says David All, a Republican new-media consultant. "You have an entire generation of folks under age 25 no longer using e-mails, not even using Facebook; a majority are using text messaging," All says. "I get Obama's text messages, and every one is exactly what it should be. It is never pointless, it is always worth reading, and it has an action for you to take. You can have hundreds of recipients on a text message. You have hundreds of people trying to change the world in 160 characters or less. What's the SMS strategy for John McCain? None."

The generational differences between the Obama and McCain campaigns may be best symbolized by the distinctly retro "Pork Invaders," a game on the McCain site (it's also a Facebook application) styled after Space Invaders, the arcade game of the late 1970s. Pork Invaders allows you to fire bullets that say "veto" at slow-moving flying pigs and barrels.

But it's not that the campaign isn't trying to speak to the youth of today, as opposed to the youth of decades ago. Lately McCain has been having his daughter Meghan and two friends write a "bloggette" from the campaign trail. The bloggette site features a silhouette of a fetching woman in red high-heeled shoes. "It gives a hipper, younger perspective on the campaign and makes both of her parents seem hipper and younger," says Julie Germany, director of the nonpartisan Institute for Politics, Democracy, and the Internet at George Washington University. The McCain campaign did not reply to several interview requests, but Germany predicts that the campaign will exploit social networking in time to make a difference in November. "What we will see is that the McCain online campaign is using the Internet just as effectively to meet its goals as the Obama campaign," she says. Over the summer, the McCain campaign refreshed its website. But Rasiej, for one, doubts that McCain has enough time to make up lost ground.

A Networked White House? The obvious next step for MyBO is to serve as a get-out-the-vote engine in November. All campaigns scrutinize public records showing who is registered to vote and whether they have voted in past elections. The Obama campaign will be able to merge this data with MyBO data. All MyBO members' activity will have been chronicled: every house party they attended, each online connection, the date and amount of each donation. Rasiej sees how it might play out: the reliable voters who signed up on MyBO but did little else may be left alone. The most active ones will be deployed to get the unreliable voters—whether MyBO members or not—to the polls. And personalized pitches can be dished up, thanks to the MyBO database. "The more contextual information they can provide the field operation, the better turnout they will have," he says.

If Obama is elected, his Web-oriented campaign strategy could carry over into his presidency. He could encourage his supporters to deluge members of Congress with calls and e-mails, or use the Web to organize collective research on policy questions. The campaign said in one of its prepared statements that "it's certain that the relationships that have been built between Barack Obama and his supporters, and between supporters themselves, will not end on Election Day." But whether or not a President Obama takes MyBO into the West Wing, it's clear that the phenomenon will forever transform campaigning. "We're scratching the surface," Trippi says. "We're all excited because he's got one million people signed up—but we are 300 million people in this country. We are still at the infancy stages of what social-networking technologies are going to do, not just in our politics but in everything. There won't be any campaign in 2012 that doesn't try to build a social network around it."

Lessig warns that if Obama wins but doesn't govern according to principles of openness and change, as promised, supporters may not be so interested in serving as MyBO foot soldiers in 2012. "The thing they [the Obama camp] don't quite recognize is how much of their enormous support comes from the perception that this is someone different," Lessig says. "If they behave like everyone else, how much will that stanch the passion of his support?"

But for now, it's party time. At the end of June, after Clinton suspended her campaign, MyBO put out a call for the faithful to organize house parties under a "Unite for Change" theme. More than 4,000 parties were organized nationwide on June 28; I logged in and picked three parties from about a dozen in the Boston area.

My first stop was a house party in the tony suburb of Winchester, where several couples dutifully watched an Obama-supplied campaign video. Host Mary Hart, an art professor in her 50s, said that Obama and his website made her "open my house to strangers and really get something going." She added, "I'm e-mailing people I haven't seen in 20 years. We have this tremendous ability to use this technology to network with people. Why don't we use it?"

Next stop was a lawn party in the Boston neighborhood of Roxbury, whose organizer, Sachielle Samedi, 34, wore a button that said "Hot Chicks Dig Obama." She said that support for the Obama candidacy drew neighbors together. At the party, Wayne Dudley, a retired history professor, met a kindred spirit: Brian Murdoch, a 54-year-old Episcopal priest. The two men buttonholed me for several minutes; Dudley predicted that Obama would bring about "a new world order centered on people of integrity." Murdoch nodded vigorously. It was a fine MyBO moment.

My evening ended at a packed post-collegiate party in a Somerville walk-up apartment. Host Rebecca Herst, a 23-year-old program assistant with the Jewish Organizing Initiative, said that MyBO—unlike Facebook—allowed her to quickly upload her entire Gmail address book, grafting her network onto Obama's. "It will be interesting to see what develops after this party, because now I'm connected to all these people," she shouted over the growing din. Two beery young men, heading for the exits, handed her two checks for $20. Herst tucked the checks into her back pocket.

Andrew Sullivan

Why I Blog

For centuries, writers have experimented with forms that evoke the imperfection of thought, the inconstancy of human affairs, and the chastening passage of time. But as blogging evolves as a literary form, it is generating a new and quintessentially postmodern idiom that's enabling writers to express themselves in ways that have never been seen or understood before. Its truths are provisional, and its ethos collective and messy. Yet the interaction it enables between writer and reader is unprecedented, visceral, and sometimes brutal. And make no mistake: it heralds a golden era for journalism.

The word blog is a conflation of two words: Web and log. It contains in its four letters a concise and accurate self-description: it is a log of thoughts and writing posted publicly on the World Wide Web. In the monosyllabic vernacular of the Internet, Web log soon became the word blog.

This form of instant and global self-publishing, made possible by technology widely available only for the past decade or so, allows for no retroactive editing (apart from fixing minor typos or small glitches) and removes from the act of writing any considered or lengthy review. It is the spontaneous expression of instant thought—impermanent beyond even the ephemera of daily journalism. It is accountable in immediate and unavoidable ways to readers and other bloggers, and linked via hypertext to continuously multiplying references and sources. Unlike any single piece of print journalism, its borders are extremely porous and its truth inherently transitory. The consequences of this for the act of writing are still sinking in.

A ship's log owes its name to a small wooden board, often weighted with lead, that was for centuries attached to a line and thrown over the stern. The weight of the log would keep it in the same place in the water, like a provisional anchor, while the ship moved away. By measuring the length of line used up in a set period of time, mariners could calculate the speed of their journey (the rope itself was marked by equidistant "knots" for easy measurement). As a ship's voyage progressed, the course came to be marked down in a book that was called a log.

In journeys at sea that took place before radio or radar or satellites or sonar, these logs were an indispensable source for recording what actually happened. They helped navigators surmise where they were and how far they had traveled and how much longer they had to stay at sea. They provided accountability to a ship's owners and traders. They were designed to be as immune to faking as possible. Away from land, there was usually no reliable corroboration of events apart from the crew's own account in the middle of an expanse of blue and gray and green; and in long journeys, memories always blur and facts disperse. A log provided as accurate an account as could be gleaned in real time.

As you read a log, you have the curious sense of moving backward in time as you move forward in pages—the opposite of a book. As you piece together a narrative that was never intended as one, it seems—and is—more truthful. Logs, in this sense, were a form of human self-correction. They amended for hindsight, for the ways in which human beings order and tidy and construct the story of their lives as they look back on them. Logs require a letting-go of narrative because they do not allow for a knowledge of the ending. So they have plot as well as dramatic irony—the reader will know the ending before the writer did.

Anyone who has blogged his thoughts for an extended time will recognize this world. We bloggers have scant opportunity to collect our thoughts, to wait until events have settled and a clear pattern emerges. We blog now—as news reaches us, as facts emerge. This is partly true for all journalism, which is, as its etymology suggests, daily writing, always subject to subsequent revision. And a good columnist will adjust position and judgment and even political loyalty over time, depending on events. But a blog is not so much daily writing as hourly writing. And with that level of timeliness, the provisionality of every word is even more pressing—and the risk of error or the thrill of prescience that much greater.

No columnist or reporter or novelist will have his minute shifts or constant small contradictions exposed as mercilessly as a blogger's are. A columnist can ignore or duck a subject less noticeably than a blogger committing thoughts to pixels several times a day. A reporter can wait—must wait—until every source has confirmed. A novelist can spend months or years before committing words to the world. For bloggers, the deadline is always now. Blogging is therefore to writing what extreme sports are to athletics: more free-form, more accident-prone, less formal, more alive. It is, in many ways, writing out loud.

You end up writing about yourself, since you are a relatively fixed point in this constant interaction with the ideas and facts of the exterior world. And in this sense, the historic form closest to blogs is the diary. But with this difference: a diary is almost always a private matter. Its raw honesty, its dedication to marking life as it happens and remembering life as it was, makes it a terrestrial log. A few diaries are meant to be read by others, of course, just as correspondence could be—but usually posthumously, or as a way to compile facts for a more considered autobiographical rendering. But a blog, unlike a diary, is instantly public. It transforms this most personal and retrospective of forms into a painfully public and immediate one. It combines the confessional genre with the log form and exposes the author in a manner no author has ever been exposed before.

I remember first grappling with what to put on my blog. It was the spring of 2000 and, like many a freelance writer at the time, I had some vague notion that I needed to have a presence "online." I had no clear idea of what to do, but a friend who ran a Web-design company offered to create a site for me, and, since I was technologically clueless, he also agreed to post various essays and columns as I wrote them. Before too long, this became a chore for him, and he called me one day to say he'd found an online platform that was so simple I could henceforth post all my writing myself. The platform was called Blogger.

As I used it to post columns or links to books or old essays, it occurred to me that I could also post new writing—writing that could even be exclusive to the blog. But what? Like any new form, blogging did not start from nothing. It evolved from various journalistic traditions. In my case, I drew on my mainstream-media experience to navigate the virgin sea. I had a few early inspirations: the old Notebook section of *The New Republic,* a magazine that, under the editorial guidance of Michael Kinsley, had introduced a more English style of crisp, short commentary into what had been a more high-minded genre of American opinion writing. *The New Republic* had also pioneered a Diarist feature on the last page, which was designed to be a more personal, essayistic, first-person form of journalism. Mixing the two genres, I did what I had been trained to do—and improvised.

I'd previously written online as well, contributing to a list-serv for gay writers and helping Kinsley initiate a more discursive form of online writing for *Slate,* the first magazine published exclusively on the Web. As soon as I began writing this way, I realized that the online form rewarded a colloquial, unfinished tone. In one of my early Kinsley-guided experiments, he urged me not to think too hard before writing. So I wrote as I'd write an e-mail—with only a mite more circumspection. This is hazardous, of course, as anyone who has ever clicked Send in a fit of anger or hurt will testify. But blogging requires an embrace of such hazards, a willingness to fall off the trapeze rather than fail to make the leap.

From the first few days of using the form, I was hooked. The simple experience of being able to directly broadcast my own words to readers was an exhilarating literary liberation. Unlike the current generation of writers, who have only ever blogged, I knew firsthand what the alternative meant. I'd edited a weekly print magazine, *The New Republic,* for five years, and written countless columns and essays for a variety of traditional outlets. And in all this, I'd often chafed, as most writers do, at the endless delays, revisions, office politics, editorial fights, and last-minute cuts for space that dead-tree publishing entails. Blogging—even to an audience of a few hundred in the early days—was intoxicatingly free in comparison. Like taking a narcotic.

It was obvious from the start that it was revolutionary. Every writer since the printing press has longed for a means to publish himself and reach—instantly—any reader on Earth. Every professional writer has paid some dues waiting for an editor's nod, or enduring a publisher's incompetence, or being ground to literary dust by a legion of fact-checkers and copy editors. If you added up the time a writer once had to spend finding an outlet, impressing editors, sucking up to proprietors, and proofreading edits, you'd find another lifetime buried in the interstices. But with one click of the Publish Now button, all these troubles evaporated.

Alas, as I soon discovered, this sudden freedom from above was immediately replaced by insurrection from below. Within minutes of my posting something, even in the earliest days, readers responded. E-mail seemed to unleash their inner beast. They were more brutal than any editor, more persnickety than any copy editor, and more emotionally unstable than any colleague.

Again, it's hard to overrate how different this is. Writers can be sensitive, vain souls, requiring gentle nurturing from editors, and oddly susceptible to the blows delivered by reviewers. They survive, for the most part, but the thinness of their skins is legendary. Moreover, before the blogosphere, reporters and columnists were largely shielded from this kind of direct hazing. Yes, letters to the editor would arrive in due course and subscriptions would be canceled. But reporters and columnists tended to operate in a relative sanctuary, answerable mainly to their editors, not readers. For a long time, columns were essentially monologues published to applause, muffled murmurs, silence, or a distant heckle. I'd gotten blowback from pieces before—but in an amorphous, time-delayed, distant way. Now the feedback was instant, personal, and brutal.

And so blogging found its own answer to the defensive counterblast from the journalistic establishment. To the charges of inaccuracy and unprofessionalism, bloggers could point to the fierce, immediate scrutiny of their readers. Unlike newspapers, which would eventually publish corrections in a box of printed spinach far from the original error, bloggers had to walk the walk of self-correction in the same space and in the same format as the original screwup. The form was more accountable, not less, because there is nothing more conducive to professionalism than being publicly humiliated for sloppiness. Of course, a blogger could ignore an error or simply refuse to acknowledge mistakes. But if he persisted, he would be razzed by competitors and assailed by commenters and abandoned by readers. In an era when the traditional media found itself beset by scandals as disparate as Stephen Glass, Jayson Blair, and Dan Rather, bloggers survived the first assault on their worth. In time, in fact, the high standards expected of well-trafficked bloggers spilled over into greater accountability, transparency, and punctiliousness among the media powers that were. Even *New York Times* columnists were forced to admit when they had been wrong.

The blog remained a *superficial* medium, of course. By superficial, I mean simply that blogging rewards brevity and immediacy. No one wants to read a 9,000-word treatise online. On the Web, one-sentence links are as legitimate as thousand-word diatribes—in fact, they are often valued more. And, as Matt Drudge told me when I sought advice from the master in 2001, the key to understanding a blog is to realize that it's a broadcast, not a publication. If it stops moving, it dies. If it stops paddling, it sinks.

But the superficiality masked considerable depth—greater depth, from one perspective, than the traditional media could offer. The reason was a single technological innovation: the hyperlink. An old-school columnist can write 800 brilliant words analyzing or commenting on, say, a new think-tank report or scientific survey. But in reading it on paper, you have to take the columnist's presentation of the material on faith, or be convinced by a brief quotation (which can always be misleading out of context). Online, a hyperlink to the original source transforms the experience. Yes, a few sentences of bloggy spin may not be as satisfying as a full column, but the ability to read the primary material instantly—in as careful or shallow a fashion as you choose—can add much greater context than anything on paper. Even a blogger's chosen pull quote, unlike a columnist's, can be effortlessly checked against the original. Now this innovation, pre-dating blogs but popularized by them, is increasingly central to mainstream journalism.

A blog, therefore, bobs on the surface of the ocean but has its anchorage in waters deeper than those print media is technologically able to exploit. It disempowers the writer to that extent, of course. The blogger can get away with less and afford fewer pretensions of authority. He is—more than any writer of the past—a node among other nodes, connected but unfinished without the links and the comments and the trackbacks that make the blogosphere, at its best, a conversation, rather than a production.

A writer fully aware of and at ease with the provisionality of his own work is nothing new. For centuries, writers have experimented with forms that suggest the imperfection of human thought, the inconstancy of human affairs, and the humbling, chastening passage of time. If you compare the meandering, questioning, unresolved dialogues of Plato with the definitive, logical treatises of Aristotle, you see the difference between a skeptic's spirit translated into writing and a spirit that seeks to bring some finality to the argument. Perhaps the greatest single piece of Christian apologetics, Pascal's *Pensées,* is a series of meandering, short, and incomplete stabs at arguments, observations, insights. Their lack of finish is what makes them so compelling—arguably more compelling than a polished treatise by Aquinas.

Or take the brilliant polemics of Karl Kraus, the publisher of and main writer for *Die Fackel,* who delighted in constantly twitting authority with slashing aphorisms and rapid-fire bursts of invective. Kraus had something rare in his day: the financial wherewithal to self-publish. It gave him a fearlessness that is now available to anyone who can afford a computer and an Internet connection.

But perhaps the quintessential blogger *avant la lettre* was Montaigne. His essays were published in three major editions, each one longer and more complex than the previous. A passionate skeptic, Montaigne amended, added to, and amplified the essays for each edition, making them three-dimensional through time. In the best modern translations, each essay is annotated, sentence by sentence, paragraph by paragraph, by small letters (A, B, and C) for each major edition, helping the reader see how each rewrite added to or subverted, emphasized or ironized, the version before. Montaigne was living his skepticism, daring to show how a writer evolves, changes his mind, learns new things, shifts perspectives, grows older—and that this, far from being something that needs to be hidden behind a veneer of unchanging authority, can become a virtue, a new way of looking at the pretensions of authorship and text and truth. Montaigne, for good measure, also peppered his essays with myriads of what bloggers would call external links. His own thoughts are strewn with and complicated by the aphorisms and anecdotes of others. Scholars of the sources note that many of these "money quotes" were deliberately taken out of context, adding layers of irony to writing that was already saturated in empirical doubt.

To blog is therefore to let go of your writing in a way, to hold it at arm's length, open it to scrutiny, allow it to float in the ether for a while, and to let others, as Montaigne did, pivot you toward relative truth. A blogger will notice this almost immediately upon starting. Some e-mailers, unsurprisingly, know more about a subject than the blogger does. They will send links, stories, and facts, challenging the blogger's view of the world, sometimes outright refuting it, but more frequently adding context and nuance and complexity to an idea. The role of a blogger is not to defend against this but to embrace it.

He is similar in this way to the host of a dinner party. He can provoke discussion or take a position, even passionately, but he also must create an atmosphere in which others want to participate.

That atmosphere will inevitably be formed by the blogger's personality. The blogosphere may, in fact, be the least veiled of any forum in which a writer dares to express himself. Even the most careful and self-aware blogger will reveal more about himself than he wants to in a few unguarded sentences and publish them before he has the sense to hit Delete. The wise panic that can paralyze a writer—the fear that he will be exposed, undone, humiliated—is not available to a blogger. You can't have blogger's block. You have to express yourself now, while your emotions roil, while your temper flares, while your humor lasts. You can try to hide yourself from real scrutiny, and the exposure it demands, but it's hard. And that's what makes blogging as a form stand out: it is rich in personality. The faux intimacy of the Web experience, the closeness of the e-mail and the instant message, seeps through. You feel as if you know bloggers as they go through their lives, experience the same things you are experiencing, and share the moment. When readers of my blog bump into me in person, they invariably address me as Andrew. Print readers don't do that. It's Mr. Sullivan to them.

On my blog, my readers and I experienced 9/11 together, in real time. I can look back and see not just how I responded to the event, but how I responded to it at 3:47 that afternoon. And at 9:46 that night. There is a vividness to this immediacy that cannot be rivaled by print. The same goes for the 2000 recount, the Iraq War, the revelations of Abu Ghraib, the death of John Paul II, or any of the other history-making events of the past decade. There is simply no way to write about them in real time without revealing a huge amount about yourself. And the intimate bond this creates with readers is unlike the bond that the *The Times,* say, develops with its readers through the same events. Alone in front of a computer, at any moment, are two people: a blogger and a reader. The proximity is palpable, the moment human—whatever authority a blogger has is derived not from the institution he works for but from the humanness he conveys. This is writing with emotion not just under but always breaking through the surface. It renders a writer and a reader not just connected but linked in a visceral, personal way. The only term that really describes this is *friendship.* And it is a relatively new thing to write for thousands and thousands of friends.

These friends, moreover, are an integral part of the blog itself—sources of solace, company, provocation, hurt, and correction. If I were to do an inventory of the material that appears on my blog, I'd estimate that a good third of it is reader-generated, and a good third of my time is spent absorbing readers' views, comments, and tips. Readers tell me of breaking stories, new perspectives, and counterarguments to prevailing assumptions. And this is what blogging, in turn, does to reporting. The traditional method involves a journalist searching for key sources, nurturing them, and sequestering them from his rivals. A blogger splashes gamely into a subject and dares the sources to come to him.

Some of this material—e-mails from soldiers on the front lines, from scientists explaining new research, from dissident Washington writers too scared to say what they think in their own partisan redoubts—might never have seen the light of day before the blogosphere. And some of it, of course, is dubious stuff. Bloggers can be spun and misled as easily as traditional writers—and the rigorous source assessment that good reporters do can't be done by e-mail. But you'd be surprised by what comes unsolicited into the in-box, and how helpful it often is.

Not all of it is mere information. Much of it is also opinion and scholarship, a knowledge base that exceeds the research department of any newspaper. A good blog is your own private Wikipedia. Indeed, the most pleasant surprise of blogging has been the number of people working in law or government or academia or rearing kids at home who have real literary talent and real knowledge, and who had no outlet—until now. There is a distinction here, of course, between the edited use of e-mailed sources by a careful blogger and the often mercurial cacophony on an unmediated comments section. But the truth is out there—and the miracle of e-mail allows it to come to you.

Fellow bloggers are always expanding this knowledge base. Eight years ago, the blogosphere felt like a handful of individual cranks fighting with one another. Today, it feels like a universe of cranks, with vast, pulsating readerships, fighting with one another. To the neophyte reader, or blogger, it can seem overwhelming. But there is a connection between the intimacy of the early years and the industry it has become today. And the connection is human individuality.

The pioneers of online journalism—*Slate* and Salon—are still very popular, and successful. But the more memorable stars of the Internet—even within those two sites—are all personally branded. Daily Kos, for example, is written by hundreds of bloggers, and amended by thousands of commenters. But it is named after Markos Moulitsas, who started it, and his own prose still provides a backbone to the front-page blog. The biggest news-aggregator site in the world, the Drudge Report, is named after its founder, Matt Drudge, who somehow conveys a unified sensibility through his selection of links, images, and stories. The vast, expanding universe of The Huffington Post still finds some semblance of coherence in the Cambridge-Greek twang of Arianna; the entire world of online celebrity gossip circles the drain of Perez Hilton; and the investigative journalism, reviewing, and commentary of Talking Points Memo is still tied together by the tone of Josh Marshall. Even *Slate* is unimaginable without Mickey Kaus's voice.

What endures is a human brand. Readers have encountered this phenomenon before—*I.F. Stone's Weekly* comes to mind—but not to this extent. It stems, I think, from the conversational style that blogging rewards. What you want in a conversationalist is as much character as authority. And if you think of blogging as more like talk radio or cable news than opinion magazines or daily newspapers, then this personalized emphasis is less surprising. People have a voice for radio and a face for television. For blogging, they have a sensibility.

But writing in this new form is a collective enterprise as much as it is an individual one—and the connections between bloggers are as important as the content on the blogs. The links not only drive conversation, they drive readers. The more you link, the more others will link to you, and the more traffic and readers you will get. The zero-sum game of old media—in which *Time* benefits from *Newsweek*'s decline and vice versa—becomes win-win. It's great for *Time* to be linked to by *Newsweek* and the other way round. One of the most prized statistics in the blogosphere is therefore not the total number of readers or page views, but the "authority" you get by being linked to by other blogs. It's an indication of how central you are to the online conversation of humankind.

The reason this open-source market of thinking and writing has such potential is that the always adjusting and evolving collective mind can rapidly filter out bad arguments and bad ideas. The flip side, of course, is that bloggers are also human beings. Reason is not the only fuel in the tank. In a world where no distinction is made between good traffic and bad traffic, and where emotion often rules, some will always raise their voice to dominate the conversation; others will pander shamelessly to their readers' prejudices; others will start online brawls for the fun of it. Sensationalism, dirt, and the ease of formulaic talking points always beckon. You can disappear into the partisan blogosphere and never stumble onto a site you disagree with.

But linkage mitigates this. A Democratic blog will, for example, be forced to link to Republican ones, if only to attack and mock. And it's in the interests of both camps to generate shared traffic. This encourages polarized slugfests. But online, at least you see both sides. Reading *The Nation* or *National Review* before the Internet existed allowed for more cocooning than the wide-open online sluice gates do now. If there's more incivility, there's also more fluidity. Rudeness, in any case, isn't the worst thing that can happen to a blogger. Being ignored is. Perhaps the nastiest thing one can do to a fellow blogger is to rip him apart and fail to provide a link.

A successful blog therefore has to balance itself between a writer's own take on the world and others. Some bloggers collect, or "aggregate," other bloggers' posts with dozens of quick links and minimalist opinion topspin: Glenn Reynolds at Instapundit does this for the right-of-center; Duncan Black at Eschaton does it for the left. Others are more eclectic, or aggregate links in a particular niche, or cater to a settled and knowledgeable reader base. A "blogroll" is an indicator of whom you respect enough to keep in your galaxy. For many years, I kept my reading and linking habits to a relatively small coterie of fellow political bloggers. In today's blogosphere, to do this is to embrace marginality. I've since added links to religious blogs and literary ones and scientific ones and just plain weird ones. As the blogosphere has expanded beyond anyone's capacity to absorb it, I've needed an assistant and interns to scour the Web for links and stories and photographs to respond to and think about. It's a difficult balance, between your own interests and obsessions, and the knowledge, insight, and wit of others—but an immensely rich one. There are times, in fact, when a blogger feels less like a writer than an online disc jockey, mixing samples of tunes and generating new melodies through mashups while also making his own music. He is both artist and producer—and the beat always goes on.

If all this sounds postmodern, that's because it is. And blogging suffers from the same flaws as postmodernism: a failure to provide stable truth or a permanent perspective. A traditional writer is valued by readers precisely because they trust him to have thought long and hard about a subject, given it time to evolve in his head, and composed a piece of writing that is worth their time to read at length and to ponder. Bloggers don't do this and cannot do this—and that limits them far more than it does traditional long-form writing.

A blogger will air a variety of thoughts or facts on any subject in no particular order other than that dictated by the passing of time. A writer will instead use time, synthesizing these thoughts, ordering them, weighing which points count more than others, seeing how his views evolved in the writing process itself, and responding to an editor's perusal of a draft or two. The result is almost always more measured, more satisfying, and more enduring than a blizzard of posts. The triumphalist notion that blogging should somehow replace traditional writing is as foolish as it is pernicious. In some ways, blogging's gifts to our discourse make the skills of a good traditional writer much more valuable, not less. The torrent of blogospheric insights, ideas, and arguments places a greater premium on the person who can finally make sense of it all, turning it into something more solid, and lasting, and rewarding.

The points of this essay, for example, have appeared in shards and fragments on my blog for years. But being forced to order them in my head and think about them for a longer stretch has helped me understand them better, and perhaps express them more clearly. Each week, after a few hundred posts, I also write an actual newspaper column. It invariably turns out to be more considered, balanced, and evenhanded than the blog. But the blog will always inform and enrich the column, and often serve as a kind of free-form, free-associative research. And an essay like this will spawn discussion best handled on a blog. The conversation, in other words, is the point, and the different idioms used by the conversationalists all contribute something of value to it. And so, if the defenders of the old media once viscerally regarded blogging as some kind of threat, they are starting to see it more as a portal, and a spur.

There is, after all, something simply irreplaceable about reading a piece of writing at length on paper, in a chair or on a couch or in bed. To use an obvious analogy, jazz entered our civilization much later than composed, formal music. But it hasn't replaced it; and no jazz musician would ever claim that it could. Jazz merely demands a different way of playing and listening, just as blogging requires a different mode of writing and reading. Jazz and blogging are intimate, improvisational, and individual—but also inherently collective. And the audience talks over both.

The reason they talk while listening, and comment or link while reading, is that they understand that this is a kind of music that needs to be engaged rather than merely absorbed. To listen to jazz as one would listen to an aria is to miss the point. Reading at a monitor, at a desk, or on an iPhone provokes a querulous, impatient, distracted attitude, a demand for instant, usable information, that is simply not conducive to opening a novel or a favorite magazine on the couch. Reading on paper evokes a more relaxed and meditative response. The message dictates the medium. And each medium has its place—as long as one is not mistaken for the other.

In fact, for all the intense gloom surrounding the newspaper and magazine business, this is actually a golden era for journalism. The blogosphere has added a whole new idiom to the act of writing and has introduced an entirely new generation to nonfiction. It has enabled writers to write out loud in ways never seen or understood before. And yet it has exposed a hunger and need for traditional writing that, in the age of television's dominance, had seemed on the wane.

Words, of all sorts, have never seemed so now.

Robin McKie

Isle of Plenty

In the past 10 years, one Danish island has cut its carbon footprint by a staggering 140%. Now, with a simple grid of windfarms, solar panels and sheep, it's selling power to the mainland and taking calls from Shell.

Jorgen Tranberg looks a farmer to his roots: grubby blue overalls, crumpled T-shirt and crinkled, weather-beaten features. His laconic manner, blond hair and black clogs also reveal his Scandinavian origins. Jorgen farms at Norreskifte on Samso, a Danish island famed for its rich, sweet strawberries and delicately flavoured early potatoes. This place is steeped in history —the Vikings built ships and constructed canals here—while modern residents of Copenhagen own dozens of the island's finer houses.

But Samso has recently undergone a remarkable transformation, one that has given it an unexpected global importance and international technological standing. Although members of a tightly knit, deeply conservative community, Samsingers—with Jorgen in the vanguard—have launched a renewable-energy revolution on this windswept scrap of Scandinavia. Solar, biomass, wind and wood-chip power generators have sprouted up across the island, while traditional fossil-fuel plants have been closed and dismantled. Nor was it hard to bring about these changes. "For me, it has been a piece of cake," says Jorgen. Nevertheless, the consequences have been dramatic.

Ten years ago, islanders drew nearly all their energy from oil and petrol brought in by tankers and from coal-powered electricity transmitted to the island through a mainland cable link. Today that traffic in energy has been reversed. Samsingers now export millions of kilowatt hours of electricity from renewable energy sources to the rest of Denmark. In doing so, islanders have cut their carbon footprint by a staggering 140 per cent. And what Samso can do today, the rest of the world can achieve in the near future, it is claimed.

Last year, carbon dioxide reached a record figure of 384 parts per million—a rise of around 35 per cent on levels that existed before the Industrial Revolution. The Intergovernmental Panel on Climate Change has warned that such changes could soon have a dramatic impact on the world's weather patterns. Already, Arctic sea ice is dwindling alarmingly and scientists say the world has only a few years left to make serious carbon-output cuts before irreversible, devastating climate change ensues. Samso suggests one route for avoiding such a fate.

Everywhere you travel on the island you see signs of change. There are dozens of wind turbines of various sizes dotted across the landscape, houses have solar-panelled roofs, while a long line of giant turbines off the island's southern tip swirl in the wind. Towns are linked to district heating systems that pump hot water to homes. These are either powered by rows of solar panels covering entire fields, or by generators which burn straw from local farms, or timber chips cut from the island's woods.

None of these enterprises has been imposed by outsiders or been funded by major energy companies. Each plant is owned either by a collective of local people or by an individidual islander. The Samso revolution has been an exercise in self-determination—a process in which islanders have decided to demonstrate what can be done to alleviate climate damage while still maintaining a comfortable lifestyle.

Consider Jorgen. As he wanders round his cowsheds, he scarcely looks like an energy entrepreneur. Yet the 47-year-old farmer is a true power broker. Apart from his fields of pumpkins and potatoes, as well as his 150 cows, he has erected a giant 1 megawatt (mw) wind turbine that looms down on his 120-hectare dairy farm. Four other great machines stand beside it, swirling in Samso's relentless winds. Each device is owned either by a neighbouring farmer or by a collective of locals. In addition, Jorgen has bought a half share in an even bigger, 2.3mw generator, one of the 10 devices that guard the south coast of Samso and now help to supply a sizeable chunk of Denmark's electricity.

The people of Samso were once the producers of more than 45,000 tonnes of carbon dioxide every year—about 11 tonnes a head. Through projects like these, they have cut that figure to –15,000. (That strange minus figure comes from the fact that Samsingers export their excess wind power to mainland Denmark, where it replaces electricity that would otherwise be generated using coal or gas.) It is a remarkable transformation, wrought mainly by Samsingers themselves, albeit with the aid of some national and European Union funds and some generous, guaranteed fixed prices that Denmark provides for wind-derived electricity. The latter ensures turbines pay for themselves over a six- or seven-year period. After that, owners can expect to rake in some tidy profits.

"It has been a very good investment," admits Jorgen. "It has made my bank manager very happy. But none of us is in it just for the money. We are doing it because it is fun and it makes us feel good." Nor do his efforts stop with his turbines. Jorgen recently redesigned his cowshed so it requires little straw for bedding for his cattle. Each animal now has its own natty mattress. Instead, most of the straw from Jorgen's fields is sold to his local district heating plant, further increasing his revenue and limiting carbon dioxide production. (Carbon dioxide is absorbed as crops grow in fields. When their stalks —straw—are burned, that carbon dioxide is released, but only as a gas that has been recycled within a single growing season. By contrast, oil, coal and gas are the remains of plants that are millions of years old and so, when burned, release carbon dioxide that had been sequestered aeons ago.)

Samso's transformation owes its origin to a 1997 experiment by the Danish government. Four islands, Laeso, Samso, Aero and Mon, as well as the region of Thyholm in Jutland, were each asked to compete in putting up the most convincing plan to cut their carbon outputs and boost their renewable-energy generation. Samso won.

Although it lies at the heart of Denmark, the nation's fractured geography also ensures the island is one of its most awkward places to reach, surrounded as it is by the Kattegat, an inlet of the North Sea. To get to Samso from Copenhagen, you have to travel by train for a couple of hours to Kalundborg and then take one of the twice daily ferries to Samso. A total of 4,100 people live here, working on farms or in hotels and restaurants. The place is isolated and compact and ideal for an experiment in community politics and energy engineering —particularly as it is low-lying and windswept. Flags never droop on Samso.

The job of setting up the Samso experiment fell to Soren Harmensen, a former environmental studies teacher, with thinning greyish hair and an infectious enthusiasm for all things renewable. Outside his project's headquarters, at the Samso Energiakademi—a stylish, barn-like building designed to cut energy consumption to an absolute minimum—there is an old, rusting petrol pump parked on the front steps. A label on it says, simply: "No fuel. So what now, my love?" Step inside and you will find no shortage of answers to that question.

Soren is a proselytiser and proud of his island's success. However, achieving it was not an easy matter. It took endless meetings to get things started. Every time there was a community issue at stake, he would arrive and preach his sermon about renewable energy and its value to the island. Slowly, the idea took hold and eventually public meetings were held purely to discuss his energy schemes. Even then, the process was erratic, with individual islanders' self-interest triggering conflicts. One Samsinger, the owner of a cement factory, proposed a nuclear plant be built on the island instead of wind turbines. He would then secure the concrete contract for the reactor, he reasoned. The plan was quietly vetoed.

"We are not hippies," says Soren. "We just want to change how we use our energy without harming the planet or without giving up the good life."

Eventually the first projects were launched, a couple of turbines on the west coast, and a district heating plant. "Nothing was achieved without talk and a great deal of community involvement," says Soren, a message he has since carried round the planet. "I visited Shropshire recently," he says. "A windfarm project there was causing a huge fuss, in particular among the three villages nearest the proposed site. The planners would soothe the objections of one village, only for the other two to get angry—so local officials would turn to them. Then the first village started to object all over again. The solution was simple, of course. Give each village a turbine, I told them. The prospect of cheap electricity would have changed everyone's minds." Needless to say, this did not happen.

On another visit—this time to Islay, off the west coast of Scotland—Soren found similar problems. "I was asked to attend a public meeting to debate the idea of turning the island into a renewable energy centre like Samso. But nearly all the speakers droned on about ideals and about climate change in general. But what people really want is to be involved themselves and to do something that can make a difference to the world. That point was entirely lost.

"Later I found that a local Islay distillery was installing a new set of boilers. Why not use the excess water to heat local homes, I suggested. That would be far too much bother, I was told. Yet that was just the kind of scheme that could kick-start a renewable-energy revolution."

Of course, there is something irritating about this Scandinavian certainty. Not every community is as cohesive as Samso's, for one thing. And it should also be noted that the island's transformation has come at a price: roughly 420m kroner—about £40m—that includes money from the Danish government, the EU, local businessmen and individual members of collectives. Thus the Samso revolution cost around £10,000 per islander, although a good chunk has come from each person's own pockets. Nevertheless, if you multiply that sum by 60m—the population of Great Britain—you get a figure of around £600bn as the cost of bringing a similar revolution to Britain. It is utterly impractical, of course—a point happily acknowledged by Soren.

"This is a pilot project to show the world what can be done. We are not suggesting everyone makes the sweeping changes that we have. People should cherry pick from what we have done in order to make modest, but still meaningful carbon emission cuts. The crucial point is that we have shown that if you want to change how we generate energy, you have to start at the community level and not impose technology on people. For example, Shell heard about what we were doing and asked to be involved—but only on condition they ended up owning the turbines. We told them to go away. We are a nation of farmers, of course. We believe in self-sufficiency."

Jesper Kjems was a freelance journalist based in Copenhagen when he and his wife came to Samso for a holiday four years ago. They fell in love with the island and moved in a few months later, although neither had jobs. Jesper started playing in a local band and met Soren Harmensen, its bassist, who sold him the Samso energy dream. Today Jesper is official spokesman for the Samso project.

Outside the town of Nordby, he showed me round its district heating project. A field has been covered with solar panels mounted to face the sun. Cold water is pumped in at one end to emerge, even on a gloomy day, as seriously hot water—around 70C—which is then piped to local houses for heating and washing. On particularly dark, sunless days, the plant switches mode: wood chips are scooped by robot crane into a furnace which heats the plant's water instead. The entire system is completely automated. "There are some living creatures involved, however," adds Jesper. "A flock of sheep is sent into the field every few days to nibble the grass before it grows long enough to prevent the sun's rays hitting the panels."

Everywhere you go, you find renewable-energy enthusiasts like Jesper. Crucially, most of them are recent recruits to the cause. Nor do planning rows concerning the sight of "eyesore" wind turbines affect Samsingers as they do Britons. "No one minds wind turbines on Samso for the simple reason that we all own a share of one," says electrician Brian Kjar.

And that is the real lesson from Samso. What has happened here is a social not a technological revolution. Indeed, it was a specific requirement of the scheme, when established, that only existing, off-the-shelf renewable technology be used. The real changes have been those in attitude. Brian's house near the southern town of Orby reveals the consequences. He has his own wind turbine, which he bought second-hand for £16,000—about a fifth of its original price. This produces more electricity than his household needs, so he uses the excess to heat water that he keeps in a huge insulated tank that he also built himself. On Samso's occasional windless days, this provides heating for his home when the 70ft turbine outside his house is not moving.

"Everyone knows someone who is interested in renewable energy today," he adds. "Something like this starts with a few people. It just needs time to spread. That is the real lesson of Samso."

Dalton Conley

Rich Man's Burden

Time and money in the information age.

For many American professionals, the Labor Day holiday yesterday probably wasn't as relaxing as they had hoped. They didn't go into the office, but they were still working. As much as they may truly have wanted to focus on time with their children, their spouses or their friends, they were unable to turn off their BlackBerrys, their laptops and their work-oriented brains.

Americans working on holidays is not a new phenomenon: we have long been an industrious folk. A hundred years ago the German sociologist Max Weber described what he called the Protestant ethic. This was a religious imperative to work hard, spend little and find a calling in order to achieve spiritual assurance that one is among the saved.

Weber claimed that this ethic could be found in its most highly evolved form in the United States, where it was embodied by aphorisms like Ben Franklin's "Industry gives comfort and plenty and respect." The Protestant ethic is so deeply engrained in our culture you don't need to be Protestant to embody it. You don't even need to be religious.

But what's different from Weber's era is that it is now the rich who are the most stressed out and the most likely to be

/*New York Times*/

working the most. Perhaps for the first time since we've kept track of such things, higher-income folks work more hours than lower-wage earners do. Since 1980, the number of men in the bottom fifth of the income ladder who work long hours (over 49 hours per week) has dropped by half, according to a study by the economists Peter Kuhn and Fernando Lozano. But among the top fifth of earners, long weeks have increased by 80 percent.

This is a stunning moment in economic history: At one time we worked hard so that someday we (or our children) wouldn't have to. Today, the more we earn, the more we work, since the opportunity cost of not working is all the greater (and since the higher we go, the more relatively deprived we feel).

In other words, when we get a raise, instead of using that hard-won money to buy "the good life," we feel even more pressure to work since the shadow costs of not working are all the greater.

One result is that even with the same work hours and household duties, women with higher incomes report feeling more stressed than women with lower incomes, according to a recent study by the economists Daniel Hamermesh and Jungmin Lee. In other words, not only does more money not solve our problems at home, it may even make things worse.

It would be easy to simply lay the blame for this state of affairs on the laptops and mobile phones that litter the lives of upper-income professionals. But the truth is that technology both creates and reflects economic realities. Instead, less visible forces have given birth to this state of affairs.

One of these forces is America's income inequality, which has steadily increased since 1969. We typically think of this process as one in which the rich get richer and the poor get poorer. Surely, that should, if anything, make upper income earners able to relax.

But it turns out that the growing disparity is really between the middle and the top. If we divided the American population in half, we would find that those in the lower half have been pretty stable over the last few decades in terms of their incomes relative to one another. However, the top half has been stretching out like taffy. In fact, as we move up the ladder the rungs get spaced farther and farther apart.

The result of this high and rising inequality is what I call an "economic red shift." Like the shift in the light spectrum caused by the galaxies rushing away, those Americans who are in the top half of the income distribution experience a sensation that, while they may be pulling away from the bottom half, they are also being left further and further behind by those just above them.

And since inequality rises exponentially the higher you climb the economic ladder, the better off you are in absolute terms, the more relatively deprived you may feel. In fact, a poll of New Yorkers found that those who earned more than $200,000 a year were the most likely of any income group to agree that "seeing other people with money" makes them feel poor.

Because these forces drive each other, they trap us in a vicious cycle: Rising inequality causes us to work more to keep up in an economy increasingly dominated by status goods. That further widens income differences.

The BlackBerrys and other wireless devices that make up our portable offices facilitate this socio-economic madness, but don't cause it. So, if you are someone who is pretty well off but couldn't stop working yesterday nonetheless, don't blame your iPhone or laptop. Blame a new wrinkle in something much more antiquated: inequality.

Nicholas Carr

Is Google Making Us Stupid?

What the Internet is doing to our brains.

"Dave, stop. Stop, will you? Stop, Dave. Will you stop, Dave?" So the supercomputer HAL pleads with the implacable astronaut Dave Bowman in a famous and weirdly poignant scene toward the end of Stanley Kubrick's *2001: A Space Odyssey*. Bowman, having nearly been sent to a deep-space death by the malfunctioning machine, is calmly, coldly disconnecting the memory circuits that control its artificial brain. "Dave, my mind is going," HAL says, forlornly. "I can feel it. I can feel it."

I can feel it, too. Over the past few years I've had an uncomfortable sense that someone, or something, has been tinkering with my brain, remapping the neural circuitry, reprogramming the memory. My mind isn't going—so far as I can tell—but it's changing. I'm not thinking the way I used to think. I can feel it most strongly when I'm reading. Immersing myself in a book or a lengthy article used to be easy. My mind would get caught up in the narrative or the turns of the argument, and I'd spend hours strolling through long stretches of prose. That's rarely the case anymore. Now my concentration often starts to drift after two or three pages. I get fidgety, lose the thread, begin looking for something else to do. I feel as if I'm always dragging my wayward brain back to the text. The deep reading that used to come naturally has become a struggle.

I think I know what's going on. For more than a decade now, I've been spending a lot of time online, searching and surfing and sometimes adding to the great databases of the Internet. The Web has been a godsend to me as a writer. Research that once required days in the stacks or periodical rooms of libraries can now be done in minutes. A few Google searches, some quick clicks on hyperlinks, and I've got the telltale fact or pithy quote I was after. Even when I'm not working, I'm as likely as not to be foraging in the Web's info-thickets—reading and writing e-mails, scanning headlines and blog posts, watching videos and listening to podcasts, or just tripping from link to link to link. (Unlike footnotes, to which they're sometimes likened, hyperlinks don't merely point to related works; they propel you toward them.)

For me, as for others, the Net is becoming a universal medium, the conduit for most of the information that flows through my eyes and ears and into my mind. The advantages of having immediate access to such an incredibly rich store of information are many, and they've been widely described and duly applauded. "The perfect recall of silicon memory," *Wired*'s Clive Thompson has written, "can be an enormous boon to thinking." But that boon comes at a price. As the media theorist Marshall McLuhan pointed out in the 1960s, media are not just passive channels of information. They supply the stuff of thought, but they also shape the process of thought. And what the Net seems to be doing is chipping away my capacity for concentration and contemplation. My mind now expects to take in information the way the Net distributes it: in a swiftly moving stream of particles. Once I was a scuba diver in the sea of words. Now I zip along the surface like a guy on a Jet Ski.

I'm not the only one. When I mention my troubles with reading to friends and acquaintances—literary types, most of them—many say they're having similar experiences. The more they use the Web, the more they have to fight to stay focused on long pieces of writing. Some of the bloggers I follow have also begun mentioning the phenomenon. Scott Karp, who writes a blog about online media, recently confessed that he has stopped reading books altogether. "I was a lit major in college, and used to be [a] voracious book reader," he wrote. "What happened?" He speculates on the answer: "What if I do all my reading on the web not so much because the way I read has changed, i.e. I'm just seeking convenience, but because the way I THINK has changed?"

Bruce Friedman, who blogs regularly about the use of computers in medicine, also has described how the Internet has altered his mental habits. "I now have almost totally lost the ability to read and absorb a longish article on the web or in print," he wrote earlier this year. A pathologist who has long been on the faculty of the University of Michigan Medical School, Friedman elaborated on his comment in a telephone conversation with me. His thinking, he said, has taken on a "staccato" quality, reflecting the way he quickly scans short passages of text from many sources online. "I can't read *War and Peace* anymore," he admitted. "I've lost the ability to do that. Even a blog post of more than three or four paragraphs is too much to absorb. I skim it."

Anecdotes alone don't prove much. And we still await the long-term neurological and psychological experiments that will provide a definitive picture of how Internet use affects cognition. But a recently published study of online research habits, conducted by scholars from University College London, suggests that we may well be in the midst of a sea change in the way we read and think. As part of the five-year research program, the scholars examined computer logs documenting the behavior of visitors to two popular research sites, one operated by the British Library and one by a U.K. educational consortium, that provide access to journal articles, e-books, and other sources of written information. They found that people using the sites exhibited "a form of skimming activity," hopping from one source to another and rarely returning to any source they'd already visited. They typically read no more than one or two pages of an article or book before they would "bounce" out to another site. Sometimes they'd save a long article, but there's no evidence that they ever went back and actually read it. The authors of the study report:

> It is clear that users are not reading online in the traditional sense; indeed there are signs that new forms of "reading" are emerging as users "power browse" horizontally through titles, contents pages and abstracts going for quick wins. It almost seems that they go online to avoid reading in the traditional sense.

Thanks to the ubiquity of text on the Internet, not to mention the popularity of text-messaging on cell phones, we may well be reading more today than we did in the 1970s or 1980s, when television was our medium of choice. But it's a different kind of reading, and behind it lies a different kind of thinking—perhaps even a new sense of the self. "We are not only *what* we read," says Maryanne Wolf, a developmental psychologist at Tufts University and the author of *Proust and the Squid: The Story and Science of the Reading Brain*. "We are *how* we read." Wolf worries that the style of reading promoted by the Net, a style that puts "efficiency" and "immediacy" above all else, may be weakening our capacity for the kind of deep reading that emerged when an earlier technology, the printing press, made long and complex works of prose commonplace. When we read online, she says, we tend to become "mere decoders of information." Our ability to interpret text, to make the rich mental connections that form when we read deeply and without distraction, remains largely disengaged.

Reading, explains Wolf, is not an instinctive skill for human beings. It's not etched into our genes the way speech is. We have to teach our minds how to translate the symbolic characters we see into the language we understand. And the media or other technologies we use in learning and practicing the craft of reading play an important part in shaping the neural circuits inside our brains. Experiments demonstrate that readers of ideograms, such as the Chinese, develop a mental circuitry for reading that is very different from the circuitry found in those of us whose written language employs an alphabet. The variations extend across many regions of the brain, including those that govern such essential cognitive functions as memory and the interpretation of visual and auditory stimuli. We can expect as well that the circuits woven by our use of the Net will be different from those woven by our reading of books and other printed works.

Sometime in 1882, Friedrich Nietzsche bought a typewriter —a Malling-Hansen Writing Ball, to be precise. His vision was failing, and keeping his eyes focused on a page had become exhausting and painful, often bringing on crushing headaches. He had been forced to curtail his writing, and he feared that he would soon have to give it up. The typewriter rescued him, at least for a time. Once he had mastered touch-typing, he was able to write with his eyes closed, using only the tips of his fingers. Words could once again flow from his mind to the page.

But the machine had a subtler effect on his work. One of Nietzsche's friends, a composer, noticed a change in the style of his writing. His already terse prose had become even tighter, more telegraphic. "Perhaps you will through this instrument even take to a new idiom," the friend wrote in a letter, noting that, in his own work, his "'thoughts' in music and language often depend on the quality of pen and paper."

"You are right," Nietzsche replied, "our writing equipment takes part in the forming of our thoughts." Under the sway of the machine, writes the German media scholar Friedrich A. Kittler, Nietzsche's prose "changed from arguments to aphorisms, from thoughts to puns, from rhetoric to telegram style."

The human brain is almost infinitely malleable. People used to think that our mental meshwork, the dense connections formed among the 100 billion or so neurons inside our skulls, was largely fixed by the time we reached adulthood. But brain researchers have discovered that that's not the case. James Olds, a professor of neuroscience who directs the Krasnow Institute for Advanced Study at George Mason University, says that even the adult mind "is very plastic." Nerve cells routinely break old connections and form new ones. "The brain," according to Olds, "has the ability to reprogram itself on the fly, altering the way it functions."

As we use what the sociologist Daniel Bell has called our "intellectual technologies"—the tools that extend our mental rather than our physical capacities—we inevitably begin to take on the qualities of those technologies. The mechanical clock, which came into common use in the 14th century, provides a compelling example. In *Technics and Civilization,* the historian and cultural critic Lewis Mumford described how the clock "disassociated time from human events and helped create the belief in an independent world of mathematically measurable sequences." The "abstract framework of divided time" became "the point of reference for both action and thought."

The clock's methodical ticking helped bring into being the scientific mind and the scientific man. But it also took something away. As the late MIT computer scientist Joseph Weizenbaum observed in his 1976 book, *Computer Power and Human Reason: From Judgment to Calculation,* the conception of the world that emerged from the widespread use of timekeeping instruments "remains an impoverished version of the older one, for it rests on a rejection of those direct experiences that formed the basis for, and indeed constituted, the old reality." In deciding when to eat, to work, to sleep, to rise, we stopped listening to our senses and started obeying the clock.

The process of adapting to new intellectual technologies is reflected in the changing metaphors we use to explain ourselves to ourselves. When the mechanical clock arrived, people began thinking of their brains as operating "like clockwork." Today, in the age of software, we have come to think of them as operating "like computers." But the changes, neuroscience tells us, go much deeper than metaphor. Thanks to our brain's plasticity, the adaptation occurs also at a biological level.

The Internet promises to have particularly far-reaching effects on cognition. In a paper published in 1936, the British mathematician Alan Turing proved that a digital computer, which at the time existed only as a theoretical machine, could be programmed to perform the function of any other information-processing device. And that's what we're seeing today. The Internet, an immeasurably powerful computing system, is subsuming most of our other intellectual technologies. It's becoming our map and our clock, our printing press and our typewriter, our calculator and our telephone, and our radio and TV.

When the Net absorbs a medium, that medium is re-created in the Net's image. It injects the medium's content with hyperlinks, blinking ads, and other digital gewgaws, and it surrounds the content with the content of all the other media it has absorbed. A new e-mail message, for instance, may announce its arrival as we're glancing over the latest headlines at a newspaper's site. The result is to scatter our attention and diffuse our concentration.

The Net's influence doesn't end at the edges of a computer screen, either. As people's minds become attuned to the crazy quilt of Internet media, traditional media have to adapt to the audience's new expectations. Television programs add text crawls and pop-up ads, and magazines and newspapers shorten their articles, introduce capsule summaries, and crowd their pages with easy-to-browse info-snippets. When, in March of this year, *The New York Times* decided to devote the second and third pages of every edition to article abstracts, its design director, Tom Bodkin, explained that the "shortcuts" would give harried readers a quick "taste" of the day's news, sparing them the "less efficient" method of actually turning the pages and reading the articles. Old media have little choice but to play by the new-media rules.

Never has a communications system played so many roles in our lives—or exerted such broad influence over our thoughts—as the Internet does today. Yet, for all that's been written about the Net, there's been little consideration of how, exactly, it's reprogramming us. The Net's intellectual ethic remains obscure.

About the same time that Nietzsche started using his typewriter, an earnest young man named Frederick Winslow Taylor carried a stopwatch into the Midvale Steel plant in Philadelphia and began a historic series of experiments aimed at improving the efficiency of the plant's machinists. With the approval of Midvale's owners, he recruited a group of factory hands, set them to work on various metalworking machines, and recorded and timed their every movement as well as the operations of the machines. By breaking down every job into a sequence of small, discrete steps and then testing different ways of performing each one, Taylor created a set of precise instructions—an "algorithm," we might say today—for how each worker should work. Midvale's employees grumbled about the strict new regime, claiming that it turned them into little more than automatons, but the factory's productivity soared.

More than a hundred years after the invention of the steam engine, the Industrial Revolution had at last found its philosophy and its philosopher. Taylor's tight industrial choreography—his "system," as he liked to call it—was embraced by manufacturers throughout the country and, in time, around the world. Seeking maximum speed, maximum efficiency, and maximum output, factory owners used time-and-motion studies to organize their work and configure the jobs of their workers. The goal, as Taylor defined it in his celebrated 1911 treatise, *The Principles of Scientific Management,* was to identify and adopt, for every job, the "one best method" of work and thereby to effect "the gradual substitution of science for rule of thumb throughout the mechanic arts." Once his system was applied to all acts of manual labor, Taylor assured his followers, it would bring about a restructuring not only of industry but of society, creating a utopia of perfect efficiency. "In the past the man has been first," he declared; "in the future the system must be first."

Taylor's system is still very much with us; it remains the ethic of industrial manufacturing. And now, thanks to the growing power that computer engineers and software coders wield over our intellectual lives, Taylor's ethic is beginning to govern the realm of the mind as well. The Internet is a machine designed for the efficient and automated collection, transmission, and manipulation of information, and its legions of programmers are intent on finding the "one best method" —the perfect algorithm—to carry out every mental movement of what we've come to describe as "knowledge work."

Google's headquarters, in Mountain View, California—the Googleplex—is the Internet's high church, and the religion practiced inside its walls is Taylorism. Google, says its chief executive, Eric Schmidt, is "a company that's founded around the science of measurement," and it is striving to "systematize everything" it does. Drawing on the terabytes of behavioral data it collects through its search engine and other sites, it carries out thousands of experiments a day, according to the *Harvard Business Review,* and it uses the results to refine the algorithms that increasingly control how people find information and extract meaning from it. What Taylor did for the work of the hand, Google is doing for the work of the mind.

The company has declared that its mission is "to organize the world's information and make it universally accessible and useful." It seeks to develop "the perfect search engine," which it defines as something that "understands exactly what you mean and gives you back exactly what you want." In Google's view, information is a kind of commodity, a utilitarian resource that can be mined and processed with industrial efficiency. The more pieces of information we can "access" and the faster we can extract their gist, the more productive we become as thinkers.

Where does it end? Sergey Brin and Larry Page, the gifted young men who founded Google while pursuing doctoral degrees in computer science at Stanford, speak frequently of their desire to turn their search engine into an artificial intelligence, a HAL-like machine that might be connected directly to our brains. "The ultimate search engine is something as smart as people—or smarter," Page said in a speech a few years back. "For us, working on search is a way to work on artificial intelligence." In a 2004 interview with *Newsweek,* Brin said, "Certainly if you had all the world's information directly attached to your brain, or an artificial brain that was smarter than your brain, you'd be better off." Last year, Page told a convention of scientists that Google is "really trying to build artificial intelligence and to do it on a large scale."

Such an ambition is a natural one, even an admirable one, for a pair of math whizzes with vast quantities of cash at their disposal and a small army of computer scientists in their employ. A fundamentally scientific enterprise, Google is motivated by a desire to use technology, in Eric Schmidt's words, "to solve problems that have never been solved before," and artificial intelligence is the hardest problem out there. Why wouldn't Brin and Page want to be the ones to crack it?

Still, their easy assumption that we'd all "be better off" if our brains were supplemented, or even replaced, by an artificial intelligence is unsettling. It suggests a belief that intelligence is the output of a mechanical process, a series of discrete steps that can be isolated, measured, and optimized. In Google's world, the world we enter when we go online, there's little place for the fuzziness of contemplation. Ambiguity is not an opening for insight but a bug to be fixed. The human brain is just an outdated computer that needs a faster processor and a bigger hard drive.

The idea that our minds should operate as high-speed data-processing machines is not only built into the workings of the Internet, it is the network's reigning business model as well. The faster we surf across the Web—the more links we click and pages we view—the more opportunities Google and other companies gain to collect information about us and to feed us advertisements. Most of the proprietors of the commercial Internet have a financial stake in collecting the crumbs of data we leave behind as we flit from link to link—the more crumbs, the better. The last thing these companies want is to encourage leisurely reading or slow, concentrated thought. It's in their economic interest to drive us to distraction.

Maybe I'm just a worrywart. Just as there's a tendency to glorify technological progress, there's a countertendency to expect the worst of every new tool or machine. In Plato's *Phae-*

drus, Socrates bemoaned the development of writing. He feared that, as people came to rely on the written word as a substitute for the knowledge they used to carry inside their heads, they would, in the words of one of the dialogue's characters, "cease to exercise their memory and become forgetful." And because they would be able to "receive a quantity of information without proper instruction," they would "be thought very knowledgeable when they are for the most part quite ignorant." They would be "filled with the conceit of wisdom instead of real wisdom." Socrates wasn't wrong—the new technology did often have the effects he feared—but he was shortsighted. He couldn't foresee the many ways that writing and reading would serve to spread information, spur fresh ideas, and expand human knowledge (if not wisdom).

The arrival of Gutenberg's printing press, in the 15th century, set off another round of teeth gnashing. The Italian humanist Hieronimo Squarciafico worried that the easy availability of books would lead to intellectual laziness, making men "less studious" and weakening their minds. Others argued that cheaply printed books and broadsheets would undermine religious authority, demean the work of scholars and scribes, and spread sedition and debauchery. As New York University professor Clay Shirky notes, "Most of the arguments made against the printing press were correct, even prescient." But, again, the doomsayers were unable to imagine the myriad blessings that the printed word would deliver.

So, yes, you should be skeptical of my skepticism. Perhaps those who dismiss critics of the Internet as Luddites or nostalgists will be proved correct, and from our hyperactive, data-stoked minds will spring a golden age of intellectual discovery and universal wisdom. Then again, the Net isn't the alphabet, and although it may replace the printing press, it produces something altogether different. The kind of deep reading that a sequence of printed pages promotes is valuable not just for the knowledge we acquire from the author's words but for the intellectual vibrations those words set off within our own minds. In the quiet spaces opened up by the sustained, undistracted reading of a book, or by any other act of contemplation, for that matter, we make our own associations, draw our own inferences and analogies, foster our own ideas. Deep reading, as Maryanne Wolf argues, is indistinguishable from deep thinking.

If we lose those quiet spaces, or fill them up with "content," we will sacrifice something important not only in our selves but in our culture. In a recent essay, the playwright Richard Foreman eloquently described what's at stake:

> I come from a tradition of Western culture, in which the ideal (my ideal) was the complex, dense and "cathedral-like" structure of the highly educated and articulate personality—a man or woman who carried inside themselves a personally constructed and unique version of the entire heritage of the West. [But now] I see within us all (myself included) the replacement of complex inner density with a new kind of self—evolving under the pressure of information overload and the technology of the "instantly available."

As we are drained of our "inner repertory of dense cultural inheritance," Foreman concluded, we risk turning into "'pancake people'—spread wide and thin as we connect with that vast network of information accessed by the mere touch of a button."

I'm haunted by that scene in *2001*. What makes it so poignant, and so weird, is the computer's emotional response to the disassembly of its mind: its despair as one circuit after another goes dark, its childlike pleading with the astronaut—"I can feel it. I can feel it. I'm afraid"—and its final reversion to what can only be called a state of innocence. HAL's outpouring of feeling contrasts with the emotionlessness that characterizes the human figures in the film, who go about their business with an almost robotic efficiency. Their thoughts and actions feel scripted, as if they're following the steps of an algorithm. In the world of *2001,* people have become so machinelike that the most human character turns out to be a machine. That's the essence of Kubrick's dark prophecy: as we come to rely on computers to mediate our understanding of the world, it is our own intelligence that flattens into artificial intelligence.

The Onion

Area Eccentric Reads Entire Book

GREENWOOD, IN—Sitting in a quiet downtown diner, local hospital administrator Philip Meyer looks as normal and well-adjusted as can be. Yet, there's more to this 27-year-old than first meets the eye: Meyer has recently finished reading a book.

Yes, the whole thing.

"It was great," said the peculiar Indiana native, who, despite owning a television set and having an active social life, read every single page of *To Kill a Mockingbird* by Harper Lee. "Especially the way things came together for Scout in the end. Very good."

Meyer, who never once jumped ahead to see what would happen and avoided skimming large passages of text in search of pictures, first began his oddball feat a week ago. Three days later, the eccentric Midwesterner was still at it, completing chapter after chapter, seemingly of his own free will.

"The whole thing was really engrossing," said Meyer, referring not to a movie, video game, or competitive sports match, but rather a full-length, 288-page novel filled entirely with words. "There were days when I had a hard time putting it down."

Even more bizarre, Meyer is believed to have done most

of his reading during his spare time—time when the outwardly healthy and stable resident could have literally been doing anything else, be it aimlessly surfing the Internet, taking a nap, or simply just staring at his bedroom wall.

"It'd be nice to read it again at some point," Meyer continued, as if that were a perfectly natural thing to say.

While it's difficult to imagine what compelled Meyer to read more than just the back cover of *To Kill a Mockingbird,* friends and family members claim the strange behavior goes all the way back to his childhood.

"I remember when Phil was a little kid, instead of picking up a book, getting bored, and then throwing it at his sister, he'd actually sit down and read the whole thing," said mother Susan Meyer, who declared she has long given up trying to explain her son's unusual hobby. "At the time, we thought it was just a phase he was going through. I guess we were wrong."

Over the years, Meyer has read dozens of books from beginning to end, regardless of whether he was forced to do so by a professor in school or whether a film version of the reading material already existed. According to girlfriend Jessica Kohler, he even uses a special cardboard marking device so that he can keep track of where he has stopped reading and later return to that exact same place.

"I used to find Phil's reading kind of charming because I had never really met anyone who read outside of a waiting room," Kohler said. "But more and more, it just feels odd, you know? He can't even go to the beach without bringing one of his books along."

According to behavioral psychologist Dr. Elizabeth Schulz, Meyer's reading of entire books is abnormal and may be indicative of a more serious obsession with reading.

"Instead of just zoning out during a bus ride or spending hour after hour watching YouTube videos at night, Mr. Meyer, unlike most healthy males, looks to books for gratification," Schulz said. "Really, it's a classic case of deviant behavior."

"At least, that's what it seems like from what little I've skimmed on the topic," she added.

As bizarre as it may seem, Meyer isn't alone. Once a month, he and several other Greenwood residents reportedly gather at night not only to read books all the way through, but also to discuss them at length.

"I don't know, it's like this weird 'book club' they're all a part of," said Brian Cummings, a longtime coworker and friend of Meyer's. "Seriously, what a bunch of freaks."

danah boyd

Reflections on Lori Drew, Bullying, and Strategies for Helping Kids

On October 17, 2006, 13-year-old Megan Meier committed suicide in her home in Missouri. Megan's suicide was initially attributed to malicious comments that a boy she liked, "Josh Evans," posted on MySpace. But as the story unfolded, it turned out that "Josh" did not exist. His profile had been created by Megan's former best friend, Ashley Grills, along with Ashley's mother, Lori Drew. The massive public outcry generated by this incident focused on the fact that an adult had participated in the online bullying of a minor.

The involvement of Lori Drew (an adult) in the suicide of Megan Meier has been an unavoidable topic. Last week, Drew was tried on three counts of accessing computers without authorization, a legal statute meant to stop hackers. She was acquitted of all felonies but convicted of three misdemeanors. The lawsuit itself was misdirected and clearly the result of prosecutors wanting to get her by any means possible. But in focusing on the technology, the prosecutors reinforced the problematic view that technology was responsible for this atrocity.

Let's be clear. Megan Meier's suicide is a tragedy. The fact that it was precipitated by bullying is horrific. And the fact that an adult was involved is downright heinous. But allowing the conversation about this tragedy to focus on MySpace has caused people to lose track of the core problems.

Lori Drew is effectively a "helicopter parent." She believed that Meier was bullying her daughter. She also believed that her daughter was innocent of any wrong-doing. (While there is no way to prove or disprove that latter belief, it is important for parents to understand that bullying is often reciprocal. Teens who are bullied *frequently retaliate* and the severity of their responses can escalate over time.) In any case, rather than teaching her daughter to take the high ground, Drew got involved. She helped her daughter to bully back.

Bullying is a horrific practice, but it's also pervasive in situations in which people are struggling to attain status. Backstabbing, rumor-mongering, and enticement aren't unique to teenagers. Look at any corporate office or political campaign and you'll see some pretty nasty bullying going on. The difference is that adults have upped the ante, having learned how to manipulate and hide their tracks. In other words, adults are much better equipped to do dreadful damage in their bullying than children and teens. They have experience. And that is not a good thing.

Lori Drew abused her power as a knowledgeable adult by leveraging her experience to humiliate and torment a teen girl. Put another way, Lori Drew engaged in psychological and emotional child abuse. Most definitions of "child abuse" include psychological or emotional mistreatment, but most legal statutes focus on sexual and physical abuse and neglect in part because emotional abuse is so hard to substantiate and prosecute.

But the fact that Lori Drew used technology to abuse Megan is *irrelevant*. Sure, she couldn't have said those things to Megan's face herself, but she could've hired a "Josh Evans" to do so. (How many movies have been made of boys being roped into teen girls' humiliation schemes?) Her crime had nothing to do with technology and the fact that she was charged of computer crimes is deeply misleading. She should be tried (and convicted) of psychologically abusing a child.

So why *do* we focus on the technology? Is it because it is the new thing that we don't understand? Or is it because contending with the fact that Drew abused a minor to protect her own child might force us to face our own bad habits in this regard? How many of you have done something morally or ethically problematic to protect your child? I suspect that, at the end of the day, many parents are able to step in Lori Drew's shoes and imagine getting carried away in an effort to protect their daughter from perceived injustices. Is that why we're so preoccupied with the technology? Because it's easier than scrutinizing our own ways of responding to and coping with the challenging dilemmas of child rearing?

Let's also make one thing very clear. The Lori Drew case is *not typical.* Many legislators are clamoring to make laws based on this case and one thing we know is that bad cases make bad case law. And, as in this instance, most legislation and public discourse focus on the technology rather than the damage of psychological abuse and misuse of adult power. Public attention also tends to dwell on incidents involving minors who have been abused by adults, especially strangers, even though the vast majority of bullying (and almost half of sexual solicitation) takes place between same-age minors who know each other. And for those who think that bullying is mostly online, think again. The majority of teens believe that bullying is far worse in-person or at school than online.

This is where technology *does* become relevant. Bullying probably has not increased because of the Internet, but its visibility to adults definitely has. Kids have long been bullied by peers at school without adults ever knowing. Now adults can see it. Most of them seem to conclude from this that the Internet is the culprit—but this logic is flawed and dangerous. Stifling bullying online won't make bullying go away; it'll just send it back underground. The visibility gives us an advantage. If we see it, we can work to stop it.

APPROACHES PARENTS AND SOCIETY SHOULD TAKE TO HELP CHILDREN

Parents need to be looking out for signs of bullying by their kids and by their kids' peers. Parents should be educating kids about bullying and the damage that it does. Most bullying starts out small. If parents catch it early on, they can help their kids with tactics to minimize the escalation. The Internet makes small acts of bullying much more visible. This is a digital advantage because, for the most part, parents used to learn of bullying only once it had escalated to unbearable levels. Now they are better equipped to intervene and provide guidance before this happens.

It's important to note that bullying is best curbed in childhood, when most of us first learn that saying something mean gives us power. As a parent, you should be vigilant about never saying mean things about others in front of your child. Even about politicians whom you despise. You should also make it very clear that mean words are intolerable. Set that frame early on and reinforce. If you see mean comments online, call them out, even if they're nothing more than "your dress is ugly."

Unfortunately, not all parents are actively involved in their kids' lives, and bullying is in fact heavily correlated with problems at home. It is also frequently prompted by kids' desire to get attention by any means necessary. This vicious cycle is why we need solutions that go beyond parents and kids.

What we need are the digital equivalent of street outreach workers. When I was in college, college students volunteered to do street outreach to help teens who were on the street find resources and help. They directed them to psychologists, doctors, and social workers. We need a program like this for the digital streets. We need college-aged young adults to troll digital environments, look out for teens in trouble, and support them in finding and attaining help. We need online counselors who can work with minors to address their behavioral issues without forcing them to contend with parents or bureaucracy. We need online social workers who can connect with kids and help them understand their options.

The Internet brings the public into our homes. This terrifies most adults and the fact that they are terrified prevents them from thinking about how to use technology to their advantage. Rather than focusing on the few incidents of disturbed adults reaching out to children, let's build systems that will enable trained adults to reach out to the many disturbed children who are crying out for help through online systems. We need social and governmental infrastructure to build this, but it's important. The teens who are hurting online are also hurting offline. We can silence their online cries by locking down the Internet, but it doesn't do a damn thing to address the core problem. We have the tools to do something about this. We just need the will and the want.

I wish we could turn back the clock and protect Megan Meier from the torment of Drew and her daughter. But we can't. And I'm not sure that there are any legal or technical measures that would do one drop of good in preventing a similar case—though I would be very happy to see more laws concerning the psychological abuse of minors by adults put on the books . . . not to prevent but to prosecute. What we *can* do is put structures in play to help at-risk children. Many of these children are invisible. Their plight doesn't get the broad media coverage that Megan Meier got: there are far too many of them and their stories have none of the glitz.

These are the kids who are being beaten at home and blog about it. Who publicly humiliate other kids to get attention. Who seek sex with strangers as a form of validation. Who are lonely, suicidal, and self-destructive. They are online. They are calling out for help. Why aren't we listening? And why are we blaming the messenger instead?

Joshua Davis

Secret Geek A-Team Hacks Back, Defends Worldwide Web

One man's secret effort to prevent the collapse of the Internet.

In June 2005, a balding, slightly overweight, perpetually T-shirt-clad 26-year-old computer consultant named Dan Kaminsky decided to get in shape. He began by scanning the Internet for workout tips and read that five minutes of sprinting was the equivalent of a half-hour jog. This seemed like a great shortcut—an elegant exercise hack—so he bought some running shoes at the nearest Niketown. That same afternoon, he laced up his new kicks and burst out the front door of his Seattle apartment building for his first five-minute workout. He took a few strides, slipped on a concrete ramp and crashed to the sidewalk, shattering his left elbow.

He spent the next few weeks stuck at home in a Percocet-tinged haze. Before the injury, he'd spent his days testing the inner workings of software programs. Tech companies hired him to root out security holes before hackers could find them. Kaminsky did it well. He had a knack for breaking things—bones and software alike.

But now, laid up in bed, he couldn't think clearly. His mind drifted. Running hadn't worked out so well. Should he buy a stationary bike? Maybe one of those recumbent

/*Wired*/

jobs would be best. He thought about partying in Las Vegas . . . mmm, martinis . . . and recalled a trick he'd figured out for getting free Wi-Fi at Starbucks.

As his arm healed, the details of that Starbucks hack kept nagging at him. He remembered that he had gotten into Starbucks' locked network using the domain name system, or DNS. When someone types google.com into a browser, DNS has a list of exactly where Google's servers are and directs the traffic to them. It's like directory assistance for the Internet. At Starbucks, the port for the low-bandwidth DNS connection—port 53—was left open to route customers to the *Pay for Starbucks Wi-Fi* Web page.

So, rather than pay, Kaminsky used port 53 to access the open DNS connection and get online. It was free but super-slow, and his friends mocked him mercilessly. To Kaminsky that was an irresistible challenge. After weeks of studying the minutiae of DNS and refining his hack, he was finally able to stream a 12-second animated video of Darth Vader dancing a jig with Michael Flatley. (The clip paired the Lord of the Sith with the Lord of the Dance.)

That was more than a year ago, but it still made him smile. DNS was the unglamorous underbelly of the Internet, but it had amazing powers. Kaminsky felt drawn to the obscure, often-ignored protocol all over again.

Maybe the painkillers loosened something in his mind, because as Kaminsky began to think more deeply about DNS he became convinced that something wasn't right. He couldn't quite figure it out, but the feeling stuck with him even after he stopped taking the pain pills. He returned to work full time and bought a recumbent stationary bike. He got hired to test the security of Windows Vista before it was released, repeatedly punching holes in it for Microsoft. Still, in the back of his mind, he was sure that the entire DNS system was vulnerable to attack.

Then last January, on a drizzly Sunday afternoon, he flopped down on his bed, flipped open his laptop, and started playing games with DNS. He used a software program called Scapy to fire random queries at the system. He liked to see how it would respond and decided to ask for the location of a series of nonexistent Web pages at a Fortune 500 company. Then he tried to trick his DNS server in San Diego into thinking that he knew the location of the bogus pages.

Suddenly it worked. The server accepted one of the fake pages as real. But so what? He could now supply fake information for a page nobody would ever visit. Then he realized that the server was willing to accept more information from him. Since he had supplied data about one of the company's Web pages, it believed that he was an authoritative source for *general* information about the company's domain. The server didn't know that the Web page didn't exist—it was listening to Kaminsky now, as if it had been hypnotized.

When DNS was created in 1983, it was designed to be helpful and trusting—it's directory assistance, after all. It was a time before hacker conventions and Internet banking. Plus, there were only a few hundred servers to keep track of. Today, the humble protocol stores the location of a billion Web addresses and routes every piece of Internet traffic in the world.

Security specialists have been revamping and strengthening DNS for more than two decades. But buried beneath all this tinkering, Kaminsky had just discovered a vestige of that original helpful and trusting program. He was now face-to-face with the behemoth's almost childlike core, and it was perfectly content to accept any information he wanted to supply about the location of the Fortune 500 company's servers.

Kaminsky froze. This was far more serious than anything he could have imagined. It was the ultimate hack. He was looking at an error coded into the heart of the Internet's infrastructure. This was not a security hole in Windows or a software bug in a Cisco router. This would allow him to reassign any Web address, reroute anyone's email, take over banking sites, or simply scramble the entire global system. The question was: Should he try it?

The vulnerability gave him the power to transfer millions out of bank accounts worldwide. He lived in a barren one-bedroom apartment and owned almost nothing. He rented the bed he was lying on as well as the couch and table in the living room. The walls were bare. His refrigerator generally contained little more than a few forgotten slices of processed cheese and a couple of Rockstar energy drinks. Maybe it was time to upgrade his lifestyle.

Or, for the sheer geeky joy of it, he could reroute all of .com into his laptop, the digital equivalent of channeling the Mississippi into a bathtub. It was a moment hackers around the world dream of—a tool that could give them unimaginable power. But maybe it was best simply to close his laptop and forget it. He could pretend he hadn't just stumbled over a skeleton key to the Net. Life would certainly be less complicated. If he stole money, he'd risk prison. If he told the world, he'd be the messenger of doom, potentially triggering a collapse of Web-based commerce.

But who was he kidding? He was just some guy. The problem had been coded into Internet architecture in 1983. It was 2008. Somebody must have fixed it by now. He typed a quick series of commands and pressed enter. When he tried to access the Fortune 500 company's Web site, he was redirected to an address he himself had specified.

"Oh shit," he mumbled. "I just broke the Internet."

Paul Vixie, one of the creators of the most widely used DNS software, stepped out of a conference in San Jose. A curious email had just popped up on his laptop. A guy named Kaminsky said he'd found a serious flaw in DNS and wanted to talk. He sent along his phone number.

Vixie had been working with DNS since the 1980s and had helped solve some serious problems over the years. He was president of the Internet Systems Consortium, a nonprofit that distributed BIND 9, his DNS software. At 44, he was considered the godfather of DNS. If there was a fundamental error in DNS, he probably would have fixed it long ago.

But to be on the safe side, Vixie decided to call Kaminsky. He picked up immediately and within minutes had outlined the flaw. A series of emotions swept over Vixie. What he was hearing shouldn't be possible, and yet everything the kid said was logical. By the end of the third minute, Vixie realized that Kaminsky had uncovered something that the best minds in computer science had overlooked. This affected not just BIND 9 but almost all DNS software. Vixie felt a deep flush of embarrassment, followed by a sense of pure panic.

"The first thing I want to say to you," Vixie told Kaminsky, trying to contain the flood of feeling, "is never, ever repeat what you just told me over a cell phone."

Vixie knew how easy it was to eavesdrop on a cell signal, and he had heard enough to know that he was facing a problem of global significance. If the information were intercepted by the wrong people, the wired world could be held ransom. Hackers could wreak havoc. Billions of dollars were at stake, and Vixie wasn't going to take any risks.

From that moment on, they would talk only on landlines, in person, or via heavily encrypted email. If the information in an email were accidentally copied onto a hard drive, that hard drive would have to be completely erased, Vixie said. Secrecy was critical. They had to find a solution before the problem became public.

Andreas Gustafsson knew something was seriously wrong. Vixie had emailed the 43-year-old DNS researcher in Espoo, Finland, asking to talk at 7 pm on a hardwired line. No cell phones.

Gustafsson hurried into the freezing March evening—his only landline was the fax in his office a brisk mile walk away. When he arrived, he saw that the machine didn't have a handset. Luckily, he had an analog phone lying around. He plugged it in, and soon it let off an old-fashioned metallic ring.

Gustafsson hadn't spoken to Vixie in years, but Vixie began the conversation by reading aloud a series of numbers —a code that would later allow him to authenticate Gustafsson's emails and prove that he was communicating with the right person. Gustafsson responded with his own authenticating code. With that out of the way, Vixie got to his point: *Find a flight to Seattle now.*

Wouter Wijngaards got a call as well, and the message was the same. The Dutch open source programmer took the train to the airport in Amsterdam, got on a 10-hour flight to Seattle, and arrived at the Silver Cloud Inn in Redmond, Washington, on March 29. He had traveled all the way from Europe, and he didn't even know why. Like Gustafsson, he had simply been told to show up in Building Nine on the Microsoft campus at 10 am on March 31.

In the lobby of the Silver Cloud, Wijngaards met Florian Weimer, a German DNS researcher he knew. Weimer was talking with Chad Dougherty, the DNS point man from Carnegie Mellon's Software Engineering Institute. Wijngaards joined the conversation—they were trying to figure out where to have dinner. Nobody talked about why some of the world's leading DNS experts happened to bump into one another near the front desk of this generic US hotel. Vixie had sworn each of them to secrecy. They simply went out for Vietnamese food and avoided saying anything about DNS.

The next morning, Kaminsky strode to the front of the conference room at Microsoft headquarters before Vixie could introduce him or even welcome the assembled heavy hitters. The 16 people in the room represented Cisco Systems, Microsoft, and the most important designers of modern DNS software.

Vixie was prepared to say a few words, but Kaminsky assumed that everyone was there to hear what he had to say. After all, he'd earned the spotlight. He hadn't sold the discovery to the Russian mob. He hadn't used it to take over banks. He hadn't destroyed the Internet. He was actually losing money on the whole thing: As a freelance computer consultant, he had taken time off work to save the world. In return, he deserved to bask in the glory of discovery. Maybe his name would be heralded around the world.

Kaminsky started by laying out the timeline. He had discovered a devastating flaw in DNS and would explain the details in a moment. But first he wanted the group to know that they didn't have much time. On August 6, he was going to a hacker convention in Las Vegas, where he would stand before the world and unveil his amazing discovery. If there was a solution, they'd better figure it out by then.

But did Kaminsky have the goods? DNS attacks were nothing new and were considered difficult to execute. The most practical attack—widely known as cache poisoning—required a hacker to submit data to a DNS server at the exact moment that it updated its records. If he succeeded, he could change the records. But, like sperm swimming toward an egg, whichever packet got there first—legitimate or malicious—locked everything else out. If the attacker lost the race, he would have to wait until the server updated again, a moment that might not come for days. And even if he timed it just right, the server required a 16-bit ID number. The hacker had a 1-in-65,536 chance of guessing it correctly.

It could take years to successfully compromise just one domain.

The experts watched as Kaminsky opened his laptop and connected the overhead projector. He had created a "weaponized" version of his attack on this vulnerability to demonstrate its power. A mass of data flashed onscreen and told the story. In less than 10 seconds, Kaminsky had compromised a server running BIND 9, Vixie's DNS routing software, which controls 80 percent of Internet traffic. It was undeniable proof that Kaminsky had the power to take down large swaths of the Internet.

The tension in the room rose as Kaminsky kept talking. The flaw jeopardized more than just the integrity of Web sites. It would allow an attacker to channel email as well. A hacker could redirect almost anyone's correspondence, from a single user's to everything coming and going between multinational corporations. He could quietly copy it before sending it along to its original destination. The victims would never know they had been compromised.

This had serious implications. Since many "forgot my password" buttons on banking sites rely on email to verify identity, an attacker could press the button, intercept the email, and change the password to anything he wanted. He would then have total access to that bank account.

"We're hosed," Wijngaards thought.

It got worse. Most Internet commerce transactions are encrypted. The encryption is provided by companies like VeriSign. Online vendors visit the VeriSign site and buy the encryption; customers can then be confident that their transactions are secure.

But not anymore. Kaminsky's exploit would allow an attacker to redirect VeriSign's Web traffic to an exact functioning replica of the VeriSign site. The hacker could then offer his own encryption, which, of course, he could unlock later.

Unsuspecting vendors would install the encryption and think themselves safe and ready for business. A cornerstone of secure Internet communication was in danger of being destroyed.

David Ulevitch smiled despite himself. The founder of OpenDNS, a company that operates DNS servers worldwide, was witnessing a tour de force—the geek equivalent of Michael Phelps winning his eighth gold medal. As far as Ulevitch was concerned, there had never been a vulnerability of this magnitude that was so easy to use. "This is an amazingly catastrophic attack," he marveled with a mix of grave concern and giddy awe.

It was a difficult flight back to San Francisco for Sandy Wilbourn, vice president of engineering for Nominum, a company hired by broadband providers to supply 150 million customers with DNS service. What he heard in Redmond was overwhelming—a 9 out of 10 on the scale of disasters. He might have given it a 10, but it was likely to keep getting worse. He was going to give this one some room to grow.

One of Wilbourn's immediate concerns was that about 40 percent of the country's broadband Internet ran through his servers. If word of the vulnerability leaked, hackers could quickly compromise those servers.

In his Redwood City, California, office, he isolated a hard drive so no one else in the company could access it. Then he called in his three top engineers, shut the door, and told them that what he was about to say couldn't be shared with anyone —not at home, not at the company. Even their interoffice email would have to be encrypted from now on.

Their task: Make a change to the basic functioning of Nominum's DNS servers. They and their customers would have to do it without the usual testing or feedback from outside the group. The implementation—the day the alteration went live to millions of people—would be its first real-world test.

It was a daunting task, but everyone who had been in Redmond had agreed to do the same thing. They would do it secretly, and then, all together on July 8, they would release their patches. If hackers didn't know there was a gaping DNS security hole before, they would know then. They just wouldn't know exactly what it was. Nominum and the other DNS software vendors would have to persuade their customers—Internet service providers from regional players such as Cablevision to giants like Comcast—to upgrade fast. It would be a race to get servers patched before hackers figured it out.

Though the Redmond group had agreed to act in concert, the patch—called the source port randomization solution—didn't satisfy everyone. It was only a short-term fix, turning what had been a 1-in-65,536 chance of success into a 1-in-4 billion shot.

Still, a hacker could use an automated system to flood a server with an endless stream of guesses. With a high-speed connection, a week of nonstop attacking would likely succeed. Observant network operators would see the spike in traffic and could easily block it. But, if overlooked, the attack could still work. The patch only papered over the fundamental flaw that Kaminsky had exposed.

On July 8, Nominum, Microsoft, Cisco, Sun Microsystems, Ubuntu, and Red Hat, among many others, released source port randomization patches. Wilbourn called it the largest multivendor patch in the history of the Internet. The ISPs and broadband carriers like Verizon and Comcast that had been asked to install it wanted to know what the problem was. Wilbourn told them it was extremely important that they deploy the patch, but the reason would remain a secret until Kaminsky delivered his talk in Las Vegas.

Even as Kaminsky was giving interviews about the urgency of patching to media outlets from the *Los Angeles Times* to CNET, the computer security industry rebelled. "Those of us . . . who have to advise management cannot tell our executives 'trust Dan,'" wrote one network administrator on a security mailing list. On one blog, an anonymous poster wrote this to Kaminsky: "You ask people not to speculate so your talk isn't blown but then you whore out minor details to every newspaper/magazine/publishing house so your name can go all over Google and gain five minutes of fame? This is why people hate you and wish you would work at McDonald's instead."

With a backlash building, Kaminsky decided to reach out to a few influential security experts in hopes of winning them over. He set up a conference call with Rich Mogull, founder of Securosis, a well-respected security firm; researcher Dino Dai Zovi; and Thomas Ptacek, a detractor who would later accuse Vixie and Kaminsky of forming a cabal.

The call occurred July 9. Kaminsky agreed to reveal the vulnerability if Mogull, Dai Zovi, and Ptacek would keep it secret until the Vegas talk August 6. They agreed, and Kaminsky's presentation laid it out for them. The security experts were stunned. Mogull wrote, "This is absolutely one of the most exceptional research projects I've seen." And in a blog post Ptacek wrote, "Dan's got the goods. *It's really f'ing good.*"

And then, on July 21, a complete description of the exploit appeared on the Web site of Ptacek's company. He claimed it was an accident but acknowledged that he had prepared a description of the hack so he could release it concurrently with Kaminsky. By the time he removed it, the description had traversed the Web. The DNS community had kept the secret for months. The computer security community couldn't keep it 12 days.

About a week later, an AT&T server in Texas was infiltrated using the Kaminsky method. The attacker took over google.com—when AT&T Internet subscribers in the Austin area tried to navigate to Google, they were redirected to a Google look-alike that covertly clicked ads. Whoever was behind the attack probably profited from the resulting increase in ad revenue.

Every day counted now. While Kaminsky, Vixie, and the others pleaded with network operators to install the patch, it's likely that other hacks occurred. But the beauty of the Kaminsky attack, as it was now known, was that it left little trace. A good hacker could reroute email, reset passwords, and transfer money out of accounts quickly. Banks were unlikely to announce the intrusions—online theft is bad PR. Better to just cover the victims' losses.

On August 6, hundreds of people crammed into a conference room at Caesars Palace to hear Kaminsky speak. The seats filled up quickly, leaving a scrum of spectators standing shoulder to shoulder in the back. A group of security experts had mockingly nominated Kaminsky for the Most Overhyped Bug award, and many wanted to know the truth: Was the massive patching effort justified, or was Kaminsky just an arrogant, media-hungry braggart?

While his grandmother handed out homemade Swedish lace cookies, Kaminsky took the stage wearing a black T-shirt featuring an image of Pac-Man at a dinner table. He tried for modesty. "Who am I?" he asked rhetorically. "Some guy. I do code."

The self-deprecation didn't suit him. He had the swagger of a rock star and adopted the tone of a misunderstood genius. After detailing the scope of the DNS problem, he stood defiantly in front of a bullet point summary of the attack and said, "People called BS on me. This is my reply."

By this time, hundreds of millions of Internet users were protected. The bomb had been defused. The problem was, there was little agreement on what the long-term solution should be. Most discussion centered around the concept of authenticating every bit of DNS traffic. It would mean that every computer in the world—from iPhones to corporate server arrays—would have to carry DNS authentication software. The root server could guarantee that it was communicating with the real .com name server, and .com would receive cryptological assurance that it was dealing with, say, the real Google. An impostor packet wouldn't be able to authenticate itself, putting an end to DNS attacks. The procedure is called DNSSEC and has high-profile proponents, including Vixie and the US government.

But implementing a massive and complicated protocol like DNSSEC isn't easy. Vixie has actually been trying to persuade people for years, and even he hasn't succeeded. Either way, the point might turn out to be moot. Kaminsky ended his Las Vegas talk by hinting that even darker security problems lay ahead. It was the type of grandstanding that has made him a polarizing figure in the computer security community. "There is no saving the Internet," he said. "There is postponing the inevitable for a little longer."

Then he sauntered off the stage and ate one of his grandma's cookies.

Clive Thompson

Brave New World of Digital Intimacy

On Sept. 5, 2006, Mark Zuckerberg changed the way that Facebook worked, and in the process he inspired a revolt.

Zuckerberg, a doe-eyed 24-year-old C.E.O., founded Facebook in his dorm room at Harvard two years earlier, and the site quickly amassed nine million users. By 2006, students were posting heaps of personal details onto their Facebook pages, including lists of their favorite TV shows, whether they were dating (and whom), what music they had in rotation and the various ad hoc "groups" they had joined (like "Sex and the City" Lovers). All day long, they'd post "status" notes explaining their moods—"hating Monday," "skipping class b/c i'm hung over." After each party, they'd stagger home to the dorm and upload pictures of the soused revelry, and spend the morning after commenting on how wasted everybody looked. Facebook became the de facto public commons—the way students found out what everyone around them was like and what he or she was doing.

But Zuckerberg knew Facebook had one major problem: It required a lot of active surfing on the part of its users. Sure, every day your Facebook friends would update their profiles with some new tidbits; it might even be something particu-

/New York Times Magazine/

larly juicy, like changing their relationship status to "single" when they got dumped. But unless you visited each friend's page every day, it might be days or weeks before you noticed the news, or you might miss it entirely. Browsing Facebook was like constantly poking your head into someone's room to see how she was doing. It took work and forethought. In a sense, this gave Facebook an inherent, built-in level of privacy, simply because if you had 200 friends on the site—a fairly typical number—there weren't enough hours in the day to keep tabs on every friend all the time.

"It was very primitive," Zuckerberg told me when I asked him about it last month. And so he decided to modernize. He developed something he called News Feed, a built-in service that would actively broadcast changes in a user's page to every one of his or her friends. Students would no longer need to spend their time zipping around to examine each friend's page, checking to see if there was any new information. Instead, they would just log into Facebook, and News Feed would appear: a single page that—like a social gazette from the 18th century—delivered a long list of up-to-the-minute gossip about their friends, around the clock, all in one place. "A stream of everything that's going on in their lives," as Zuckerberg put it.

When students woke up that September morning and saw News Feed, the first reaction, generally, was one of panic. Just about every little thing you changed on your page was now instantly blasted out to hundreds of friends, including potentially mortifying bits of news—*Tim and Lisa broke up; Persaud is no longer friends with Matthew*—and drunken photos someone snapped, then uploaded and tagged with names. Facebook had lost its vestigial bit of privacy. For students, it was now like being at a giant, open party filled with everyone you know, able to eavesdrop on what everyone else was saying, all the time.

"Everyone was freaking out," Ben Parr, then a junior at Northwestern University, told me recently. What particularly enraged Parr was that there wasn't any way to opt out of News Feed, to "go private" and have all your information kept quiet. He created a Facebook group demanding Zuckerberg either scrap News Feed or provide privacy options. "Facebook users really think Facebook is becoming the Big Brother of the Internet, recording every single move," a California student told *The Star-Ledger* of Newark. Another chimed in, "Frankly, I don't need to know or care that Billy broke up with Sally, and Ted has become friends with Steve." By lunchtime of the first day, 10,000 people had joined Parr's group, and by the next day it had 284,000.

Zuckerberg, surprised by the outcry, quickly made two decisions. The first was to add a privacy feature to News Feed, letting users decide what kind of information went out. But the second decision was to leave News Feed otherwise intact. He suspected that once people tried it and got over their shock, they'd like it.

He was right. Within days, the tide reversed. Students began e-mailing Zuckerberg to say that via News Feed they'd learned things they would never have otherwise discovered through random surfing around Facebook. The bits of trivia that News Feed delivered gave them more things to talk about —*Why do you hate Kiefer Sutherland?*—when they met friends face to face in class or at a party. Trends spread more quickly. When one student joined a group—proclaiming her love of Coldplay or a desire to volunteer for Greenpeace—all her friends instantly knew, and many would sign up themselves. Users' worries about their privacy seemed to vanish within days, boiled away by their excitement at being so much more connected to their friends. (Very few people stopped using Facebook, and most people kept on publishing most of their information through News Feed.) Pundits predicted that

News Feed would kill Facebook, but the opposite happened. It catalyzed a massive boom in the site's growth. A few weeks after the News Feed imbroglio, Zuckerberg opened the site to the general public (previously, only students could join), and it grew quickly; today, it has 100 million users.

When I spoke to him, Zuckerberg argued that News Feed is central to Facebook's success. "Facebook has always tried to push the envelope," he said. "And at times that means stretching people and getting them to be comfortable with things they aren't yet comfortable with. A lot of this is just social norms catching up with what technology is capable of."

In essence, Facebook users didn't think they wanted constant, up-to-the-minute updates on what other people are doing. Yet when they experienced this sort of omnipresent knowledge, they found it intriguing and addictive. Why?

Social scientists have a name for this sort of incessant online contact. They call it "ambient awareness." It is, they say, very much like being physically near someone and picking up on his mood through the little things he does—body language, sighs, stray comments—out of the corner of your eye. Facebook is no longer alone in offering this sort of interaction online. In the last year, there has been a boom in tools for "microblogging": posting frequent tiny updates on what you're doing. The phenomenon is quite different from what we normally think of as blogging, because a blog post is usually a written piece, sometimes quite long: a statement of opinion, a story, an analysis. But these new updates are something different. They're far shorter, far more frequent and less carefully considered. One of the most popular new tools is Twitter, a Web site and messaging service that allows its two-million-plus users to broadcast to their friends haiku-length updates—limited to 140 characters, as brief as a mobile-phone text message—on what they're doing. There are other services for reporting where you're traveling (Dopplr) or for quickly tossing online a stream of the pictures, videos or Web sites you're looking at (Tumblr). And there are even tools that give your location. When the new iPhone, with built-in tracking, was introduced in July, one million people began using Loopt, a piece of software that automatically tells all your friends exactly where you are.

For many people—particularly anyone over the age of 30—the idea of describing your blow-by-blow activities in such detail is absurd. Why would you subject your friends to your daily minutiae? And conversely, how much of their trivia can you absorb? The growth of ambient intimacy can seem like modern narcissism taken to a new, supermetabolic extreme—the ultimate expression of a generation of celebrity-addled youths who believe their every utterance is fascinating and ought to be shared with the world. Twitter, in particular, has been the subject of nearly relentless scorn since it went online. "Who really cares what I am doing, every hour of the day?" wondered Alex Beam, a *Boston Globe* columnist, in an essay about Twitter last month. "Even I don't care."

Indeed, many of the people I interviewed, who are among the most avid users of these "awareness" tools, admit that at first they couldn't figure out why anybody would want to do this. Ben Haley, a 39-year-old documentation specialist for a software firm who lives in Seattle, told me that when he first heard about Twitter last year from an early-adopter friend who used it, his first reaction was that it seemed silly. But a few of his friends decided to give it a try, and they urged him to sign up, too.

Each day, Haley logged on to his account, and his friends' updates would appear as a long page of one- or two-line notes. He would check and recheck the account several times a day, or even several times an hour. The updates were indeed pretty banal. One friend would post about starting to feel sick; one posted random thoughts like "I really hate it when people clip their nails on the bus"; another Twittered whenever she made a sandwich—and she made a sandwich every day. Each so-called tweet was so brief as to be virtually meaningless.

But as the days went by, something changed. Haley discovered that he was beginning to sense the rhythms of his friends' lives in a way he never had before. When one friend got sick with a virulent fever, he could tell by her Twitter updates when she was getting worse and the instant she finally turned the corner. He could see when friends were heading into hellish days at work or when they'd scored a big success. Even the daily catalog of sandwiches became oddly mesmerizing, a sort of metronomic click that he grew accustomed to seeing pop up in the middle of each day.

This is the paradox of ambient awareness. Each little update—each individual bit of social information—is insignificant on its own, even supremely mundane. But taken together, over time, the little snippets coalesce into a surprisingly sophisticated portrait of your friends' and family members' lives, like thousands of dots making a pointillist painting. This was never before possible, because in the real world, no friend would bother to call you up and detail the sandwiches she was eating. The ambient information becomes like "a type of E.S.P.," as Haley described it to me, an invisible dimension floating over everyday life.

"It's like I can distantly read everyone's mind," Haley went on to say. "I love that. I feel like I'm getting to something raw about my friends. It's like I've got this heads-up display for them." It can also lead to more real-life contact, because when one member of Haley's group decides to go out to a bar or see a band and Twitters about his plans, the others see it, and some decide to drop by—ad hoc, self-organizing socializing. And when they do socialize face to face, it feels oddly as if they've never actually been apart. They don't need to ask,

"So, what have you been up to?" because they already know. Instead, they'll begin discussing something that one of the friends Twittered that afternoon, as if picking up a conversation in the middle.

Facebook and Twitter may have pushed things into overdrive, but the idea of using communication tools as a form of "co-presence" has been around for a while. The Japanese sociologist Mizuko Ito first noticed it with mobile phones: lovers who were working in different cities would send text messages back and forth all night—tiny updates like "enjoying a glass of wine now" or "watching TV while lying on the couch." They were doing it partly because talking for hours on mobile phones isn't very comfortable (or affordable). But they also discovered that the little Ping-Ponging messages felt even more intimate than a phone call.

"It's an aggregate phenomenon," Marc Davis, a chief scientist at Yahoo and former professor of information science at the University of California at Berkeley, told me. "No message is the single-most-important message. It's sort of like when you're sitting with someone and you look over and they smile at you. You're sitting here reading the paper, and you're doing your side-by-side thing, and you just sort of let people know you're aware of them." Yet it is also why it can be extremely hard to understand the phenomenon until you've experienced it. Merely looking at a stranger's Twitter or Facebook feed isn't interesting, because it seems like blather. Follow it for a day, though, and it begins to feel like a short story; follow it for a month, and it's a novel.

You could also regard the growing popularity of online awareness as a reaction to social isolation, the modern American disconnectedness that Robert Putnam explored in his book "Bowling Alone." The mobile workforce requires people to travel more frequently for work, leaving friends and family behind, and members of the growing army of the self-employed often spend their days in solitude. Ambient intimacy becomes a way to "feel less alone," as more than one Facebook and Twitter user told me.

When I decided to try out Twitter last year, at first I didn't have anyone to follow. None of my friends were yet using the service. But while doing some Googling one day I stumbled upon the blog of Shannon Seery, a 32-year-old recruiting consultant in Florida, and I noticed that she Twittered. Her Twitter updates were pretty charming—she would often post links to camera-phone pictures of her two children or videos of herself cooking Mexican food, or broadcast her agonized cries when a flight was delayed on a business trip. So on a whim I started "following" her—as easy on Twitter as a click of the mouse—and never took her off my account. (A Twitter account can be "private," so that only invited friends can read one's tweets, or it can be public, so anyone can; Seery's was public.) When I checked in last month, I noticed that she had built up a huge number of online connections: She was now following 677 people on Twitter and another 442 on Facebook. How in God's name, I wondered, could she follow so many people? Who precisely are they? I called Seery to find out.

"I have a rule," she told me. "I either have to know who you are, or I have to know of you." That means she monitors the lives of friends, family, anyone she works with, and she'll also follow interesting people she discovers via her friends' online lives. Like many people who live online, she has wound up following a few strangers—though after a few months they no longer feel like strangers, despite the fact that she has never physically met them.

I asked Seery how she finds the time to follow so many people online. The math seemed daunting. After all, if her 1,000 online contacts each post just a couple of notes each day, that's several thousand little social pings to sift through daily.

What would it be like to get thousands of e-mail messages a day? But Seery made a point I heard from many others: awareness tools aren't as cognitively demanding as an e-mail message. E-mail is something you have to stop to open and assess. It's personal; someone is asking for 100 percent of your attention. In contrast, ambient updates are all visible on one single page in a big row, and they're not really directed at you. This makes them skimmable, like newspaper headlines; maybe you'll read them all, maybe you'll skip some. Seery estimated that she needs to spend only a small part of each hour actively reading her Twitter stream.

Yet she has, she said, become far more gregarious online. "What's really funny is that before this 'social media' stuff, I always said that I'm not the type of person who had a ton of friends," she told me. "It's so hard to make plans and have an active social life, having the type of job I have where I travel all the time and have two small kids. But it's easy to tweet all the time, to post pictures of what I'm doing, to keep social relations up." She paused for a second, before continuing: "Things like Twitter have actually given me a much bigger social circle. I know more about more people than ever before."

I realized that this is becoming true of me, too. After following Seery's Twitter stream for a year, I'm more knowledgeable about the details of her life than the lives of my two sisters in Canada, whom I talk to only once every month or so. When I called Seery, I knew that she had been struggling with a three-day migraine headache; I began the conversation by asking her how she was feeling.

Online awareness inevitably leads to a curious question: What sort of relationships are these? What does it mean to have hundreds of "friends" on Facebook? What kind of friends are they, anyway?

In 1998, the anthropologist Robin Dunbar argued that each human has a hard-wired upper limit on the number of people he or she can personally know at one time. Dunbar noticed that humans and apes both develop social bonds by engaging in some sort of grooming; apes do it by picking at and smoothing one another's fur, and humans do it with conversation. He theorized that ape and human brains could manage only a finite number of grooming relationships: unless we spend enough time doing social grooming—chitchatting, trading gossip or, for apes, picking lice—we won't really feel that we "know" someone well enough to call him a friend. Dunbar noticed that ape groups tended to top out at 55 members. Since human brains were proportionally bigger, Dunbar figured that our maximum number of social connections would be similarly larger: about 150 on average. Sure enough, psychological studies have confirmed that human groupings naturally tail off at around 150 people: the "Dunbar number," as it is known. Are people who use Facebook and Twitter increasing their Dunbar number, because they can so easily keep track of so many more people?

As I interviewed some of the most aggressively social people online—people who follow hundreds or even thousands of others—it became clear that the picture was a little more complex than this question would suggest. Many maintained that their circle of true intimates, their very close friends and family, had not become bigger. Constant online contact had made those ties immeasurably richer, but it hadn't actually increased the number of them; deep relationships are still predicated on face time, and there are only so many hours in the day for that.

But where their sociality had truly exploded was in their "weak ties"—loose acquaintances, people they knew less well. It might be someone they met at a conference, or someone from high school who recently "friended" them on Facebook, or somebody from last year's holiday party. In their pre-Internet lives, these sorts of acquaintances would have quickly faded from their attention. But when one of these far-flung people suddenly posts a personal note to your feed, it is essentially a reminder that they exist. I have noticed this effect myself. In the last few months, dozens of old work colleagues I knew from 10 years ago in Toronto have friended me on Facebook, such that I'm now suddenly reading their stray comments and updates and falling into oblique, funny conversations with them. My overall Dunbar number is thus 301: Facebook (254) + Twitter (47), double what it would be without technology. Yet only 20 are family or people I'd consider close friends. The rest are weak ties—maintained via technology.

This rapid growth of weak ties can be a very good thing. Sociologists have long found that "weak ties" greatly expand your ability to solve problems. For example, if you're looking for a job and ask your friends, they won't be much help; they're too similar to you, and thus probably won't have any leads that you don't already have yourself. Remote acquaintances will be much more useful, because they're farther afield, yet still socially intimate enough to want to help you out. Many avid Twitter users—the ones who fire off witty posts hourly and wind up with thousands of intrigued followers—explicitly milk this dynamic for all it's worth, using their large online followings as a way to quickly answer almost any question. Laura Fitton, a social-media consultant who has become a minor celebrity on Twitter—she has more than 5,300 followers—recently discovered to her horror that her accountant had made an error in filing last year's taxes. She went to Twitter, wrote a tiny note explaining her problem, and within 10 minutes her online audience had provided leads to lawyers and better accountants. Fritton joked to me that she no longer buys anything worth more than $50 without quickly checking it with her Twitter network.

"I outsource my entire life," she said. "I can solve any problem on Twitter in six minutes." (She also keeps a secondary Twitter account that is private and only for a much smaller circle of close friends and family—"My little secret," she said. It is a strategy many people told me they used: one account for their weak ties, one for their deeper relationships.)

It is also possible, though, that this profusion of weak ties can become a problem. If you're reading daily updates from hundreds of people about whom they're dating and whether they're happy, it might, some critics worry, spread your emotional energy too thin, leaving less for true intimate relationships. Psychologists have long known that people can engage in "parasocial" relationships with fictional characters, like those on TV shows or in books, or with remote celebrities we read about in magazines. Parasocial relationships can use up some of the emotional space in our Dunbar number, crowding out real-life people. Danah Boyd, a fellow at Harvard's Berkman Center for Internet and Society who has studied social media for 10 years, published a paper this spring arguing that awareness tools like News Feed might be creating a whole new class of relationships that are nearly parasocial—peripheral people in our network whose intimate details we follow closely online, even while they, like Angelina Jolie, are basically unaware we exist.

"The information we subscribe to on a feed is not the same as in a deep social relationship," Boyd told me. She has seen this herself; she has many virtual admirers that have, in essence, a parasocial relationship with her. "I've been very, very sick, lately and I write about it on Twitter and my blog, and I get all these people who are writing to me telling me ways to work around the health-care system, or they're writing saying, 'Hey, I broke my neck!' And I'm like, 'You're being very nice and trying to help me, but though you feel like you know me, you don't.'" Boyd sighed. "They can *observe* you, but it's not the same as *knowing* you."

When I spoke to Caterina Fake, a founder of Flickr (a popular photo-sharing site), she suggested an even more subtle danger: that the sheer ease of following her friends' updates online has made her occasionally lazy about actually taking the time to visit them in person. "At one point I realized I had a friend whose child I had seen, via photos on Flickr, grow from birth to 1 year old," she said. "I thought, I really should go meet her in person. But it was weird; I also felt that Flickr had satisfied that getting-to-know you satisfaction, so I didn't feel the urgency. But then I was like, Oh, that's not sufficient! I should go in person!" She has about 400 people she follows online but suspects many of those relationships are tissue-fragile. "These technologies allow you to be much more broadly friendly, but you just spread yourself much more thinly over many more people."

What is it like to never lose touch with anyone? One morning this summer at my local cafe, I overheard a young woman complaining to her friend about a recent Facebook drama. Her name is Andrea Ahan, a 27-year-old restaurant entrepreneur, and she told me that she had discovered that high-school friends were uploading old photos of her to Facebook and tagging them with her name, so they automatically appeared in searches for her.

She was aghast. "I'm like, my God, these pictures are completely hideous!" Ahan complained, while her friend looked on sympathetically and sipped her coffee. "I'm wearing all these totally awful '90s clothes. I look like crap. And I'm like, Why are you people in my life, anyway? I haven't seen you in 10 years. I don't know you anymore!" She began furiously de-tagging the pictures—removing her name, so they wouldn't show up in a search anymore.

Worse, Ahan was also confronting a common plague of Facebook: the recent ex. She had broken up with her boyfriend not long ago, but she hadn't "unfriended" him, because that felt too extreme. But soon he paired up with another young woman, and the new couple began having public conversations on Ahan's ex-boyfriend's page. One day, she noticed with alarm that the new girlfriend was quoting material Ahan had e-mailed privately to her boyfriend; she suspected he had been sharing the e-mail with his new girlfriend. It is the sort of weirdly subtle mind game that becomes possible via Facebook, and it drove Ahan nuts.

"Sometimes I think this stuff is just crazy, and everybody has got to get a life and stop obsessing over everyone's trivia and gossiping," she said.

Yet Ahan knows that she cannot simply walk away from her online life, because the people she knows online won't stop talking about her, or posting unflattering photos. She needs to stay on Facebook just to monitor what's being said about her. This is a common complaint I heard, particularly from people in their 20s who were in college when Facebook appeared and have never lived as adults without online awareness. For them, participation isn't optional. If you don't dive in, other people will define who you are. So you constantly stream your pictures, your thoughts, your relationship status and what you're doing—right now!—if only to ensure the virtual version of you is accurate, or at least the one you want to present to the world.

This is the ultimate effect of the new awareness: It brings back the dynamics of small-town life, where everybody knows your business. Young people at college are the ones to experience this most viscerally, because, with more than 90 percent of their peers using Facebook, it is especially difficult for them to opt out. Zeynep Tufekci, a sociologist at the University of Maryland, Baltimore County, who has closely studied how college-age users are reacting to the world of awareness, told me that athletes used to sneak off to parties illicitly, breaking the no-drinking rule for team members. But then camera phones and Facebook came along, with students posting photos of the drunken carousing during the party; savvy coaches could see which athletes were breaking the rules. First the athletes tried to fight back by waking up early the morning after the party in a hungover daze to detag photos of themselves so they wouldn't be searchable. But that didn't work, because the coaches sometimes viewed the pictures live, as they went online at 2 a.m. So parties simply began banning all camera phones in a last-ditch attempt to preserve privacy.

"It's just like living in a village, where it's actually hard to lie because everybody knows the truth already," Tufekci said. "The current generation is never unconnected. They're never losing touch with their friends. So we're going back to a more normal place, historically. If you look at human history, the idea that you would drift through life, going from new relation to new relation, that's very new. It's just the 20th century."

Psychologists and sociologists spent years wondering how humanity would adjust to the anonymity of life in the city, the wrenching upheavals of mobile immigrant labor—a world of lonely people ripped from their social ties. We now have precisely the opposite problem. Indeed, our modern awareness tools reverse the original conceit of the Internet. When cyberspace came along in the early '90s, it was celebrated as a place where you could reinvent your identity—become someone new.

"If anything, it's identity-constraining now," Tufekci told me. "You can't play with your identity if your audience is always checking up on you. I had a student who posted that she was downloading some Pearl Jam, and someone wrote on her wall, 'Oh, right, ha-ha—I know you, and you're not into that.'" She laughed. "You know that old cartoon? 'On the Internet, nobody knows you're a dog'? On the Internet today, everybody knows you're a dog! If you don't want people to know you're a dog, you'd better stay away from a keyboard."

Or, as Leisa Reichelt, a consultant in London who writes regularly about ambient tools, put it to me: "Can you imagine a Facebook for children in kindergarten, and they never lose touch with those kids for the rest of their lives? What's that going to do to them?" Young people today are already developing an attitude toward their privacy that is simultaneously vigilant and laissez-faire. They curate their online personas as carefully as possible, knowing that everyone is watching—but they have also learned to shrug and accept the limits of what they can control.

It is easy to become unsettled by privacy-eroding aspects of awareness tools. But there is another—quite different—result of all this incessant updating: a culture of people who know much more about themselves. Many of the avid Twitterers, Flickrers and Facebook users I interviewed described an unexpected side-effect of constant self-disclosure. The act of stopping several times a day to observe what you're feeling or thinking can become, after weeks and weeks, a sort of philosophical act. It's like the Greek dictum to "know thyself," or the therapeutic concept of mindfulness. (Indeed, the question that floats eternally at the top of Twitter's Web site—"What are you doing?"—can come to seem existentially freighted. What are you doing?) Having an audience can make the self-reflection even more acute, since, as my interviewees noted, they're trying to describe their activities in a way that is not only accurate but also interesting to others: the status update as a literary form.

Laura Fitton, the social-media consultant, argues that her

constant status updating has made her "a happier person, a calmer person" because the process of, say, describing a horrid morning at work forces her to look at it objectively. "It drags you out of your own head," she added. In an age of awareness, perhaps the person you see most clearly is yourself.

Elizabeth Kolbert

Dymaxion Man

The visions of Buckminster Fuller.

One of Buckminster Fuller's earliest inventions was a car shaped like a blimp. The car had three wheels—two up front, one in the back—and a periscope instead of a rear window. Owing to its unusual design, it could be maneuvered into a parking space nose first and could execute a hundred-and-eighty-degree turn so tightly that it would end up practically where it had started, facing the opposite direction. In Bridgeport, Connecticut, where the car was introduced in the summer of 1933, it caused such a sensation that gridlock followed, and anxious drivers implored Fuller to keep it off the streets at rush hour.

Fuller called his invention the Dymaxion Vehicle. He believed that it would not just revolutionize automaking but help bring about a wholesale reordering of modern life. Soon, Fuller thought, people would be living in standardized, prefabricated dwellings, and this, in turn, would allow them to occupy regions previously considered uninhabitable—the Arctic, the Sahara, the tops of mountains. The Dymaxion Vehicle would carry them to their new homes; it would be capable of travelling on the roughest roads and—once the technology for the requisite engines had been worked out—it would also (somehow) be able to fly. Fuller envisioned the Dymaxion taking off almost vertically, like a duck.

Fuller's schemes often had the hallucinatory quality associated with science fiction (or mental hospitals). It concerned him not in the least that things had always been done a certain way in the past. In addition to flying cars, he imagined mass-produced bathrooms that could be installed like refrigerators; underwater settlements that would be restocked by submarine; and floating communities that, along with all their inhabitants, would hover among the clouds. Most famously, he dreamed up the geodesic dome. "If you are in a shipwreck and all the boats are gone, a piano top . . . that comes along makes a fortuitous life preserver," Fuller once wrote. "But this is not to say that the best way to design a life preserver is in the form of a piano top. I think that we are clinging to a great many piano tops in accepting yesterday's fortuitous contrivings." Fuller may have spent his life inventing things, but he claimed that he was not particularly interested in inventions. He called himself a "comprehensive, anticipatory design scientist"—a "comprehensivist," for short—and believed that his task was to innovate in such a way as to benefit the greatest number of people using the least amount of resources. "My objective was humanity's comprehensive success in the universe" is how he once put it. "I could have ended up with a pair of flying slippers."

Fuller's career is the subject of a new exhibition, "Buckminster Fuller: Starting with the Universe," which opens later this month at the Whitney Museum of American Art. The exhibition traces the long, loopy arc of his career from early doodlings to plans he drew up shortly before his death, twenty-five years ago this summer. It will feature studies for several of his geodesic domes and the only surviving Dymaxion Vehicle. By staging the retrospective, the Whitney raises—or, really, one should say, re-raises—the question of Fuller's relevance. Was he an important cultural figure because he produced inventions of practical value or because he didn't?

Richard Buckminster Fuller, Jr.—Bucky, to his friends—was born on July 12, 1895, into one of New England's most venerable and, at the same time, most freethinking families. His great-great-grandfather, the Reverend Timothy Fuller, a Massachusetts delegate to the Federal Constitutional Assembly, was so outraged by the Constitution's sanctioning of slavery that he came out against ratification. His great-aunt Margaret Fuller, a friend of Emerson and Thoreau, edited the transcendentalist journal *The Dial* and later became America's first female foreign correspondent.

Growing up in Milton, Massachusetts, Bucky was a boisterous but hopelessly nearsighted child; until he was fitted with glasses, he refused to believe that the world was not blurry. Like all Fuller men, he was sent off to Harvard. Halfway through his freshman year, he withdrew his tuition money from the bank to entertain some chorus girls in Manhattan. He was expelled. The following fall, he was reinstated, only to be thrown out again. Fuller never did graduate from Harvard, or any other school. He took a job with a meatpacking firm, then joined the Navy, where he invented a winchlike device for rescuing pilots of the service's primitive airplanes. (The pilots often ended up head down, under water.)

During the First World War, Fuller married Anne Hewlett, the daughter of a prominent architect, and when the war was over he started a business with his father-in-law, manufacturing bricks out of wood shavings. Despite the general prosperity of the period, the company struggled and, in 1927, nearly bankrupt, it was bought out. At just about the same time, Anne gave birth to a daughter. With no job and a new baby to support, Fuller became depressed. One day, he was walking by Lake Michigan, thinking about, in his words, "Buckminster Fuller—life or death," when he found himself suspended several feet above the ground, surrounded by sparkling light. Time seemed to stand still, and a voice spoke to him. "You do not have the right to eliminate yourself," it said. "You do not belong to you. You belong to Universe." (In Fuller's idiosyncratic English, "universe"—capitalized—is never preceded by the definite article.) It was at this point, according to Fuller, that he decided to embark on his "lifelong experiment." The experiment's aim was nothing less than determining "what, if anything," an individual could do "on behalf of all humanity." For this study, Fuller would serve both as the researcher and as the object of inquiry. (He referred to himself as Guinea Pig B, the "B" apparently being for Bucky.) Fuller moved his wife and daughter into a tiny studio in a Chicago slum and, instead of finding a job, took to spending his days in the library, reading Gandhi and Leonardo. He began to record his own ideas, which soon filled two thousand pages. In 1928, he edited the manuscript down to fifty pages, and had it published in a booklet called "4D Time Lock," which he sent out to, among others, Vincent Astor, Bertrand Russell, and Henry Ford.

Like most of Fuller's writings, "4D Time Lock" is nearly impossible to read; its sentences, Slinky-like, stretch on and on and on. (One of his biographers observed of "4D Time Lock" that "worse prose is barely conceivable.") At its heart is a critique of the construction industry. Imagine, Fuller says, what would happen if a person, seeking to purchase an automobile, had to hire a designer, then send the plans out for bid, then show them to the bank, and then have them approved by the town council, all before work on the vehicle could begin. "Few would have the temerity to go through with it," he notes, and those who did would have to pay something like fifty thousand dollars—half a million in today's money—per car. Such a system, so obviously absurd for autos, persisted for houses, Fuller argued, because of retrograde thinking. (His own failure at peddling wood-composite bricks he cited as evidence of the construction industry's recalcitrance.) What was needed was a "New Era Home," which would be "erectable in one day, complete in every detail," and, on top of that, "drudgery-proof," with "every living appliance known to mankind, built-in."

Not coincidentally, Fuller was working to design just such a home. One plan, which never made it beyond the sketching stage, called for ultra-lightweight towers to be assembled at a central location, then transported to any spot in the world, via zeppelin. (Fuller envisioned the zeppelin crew excavating the site by dropping a small bomb.) A second, only slightly less fabulous proposal was for what Fuller came to call the Dymaxion House. The hexagonal-shaped, single-family home was to be stamped out of metal and suspended from a central mast that would contain all its wiring and plumbing. When a family moved, the Dymaxion House could be disassembled and taken along, like a bed or a table. Fuller constructed a scale model of the house, which was exhibited in 1929 at Marshall Field's as part of a display of modern furniture. But no full-size version could be produced, because many of the components, including what Fuller called a "radio-television receiver," did not yet exist. Fuller estimated that it would take a billion dollars to develop the necessary technologies. Not surprisingly, the money wasn't forthcoming.

Fuller was fond of neologisms. He coined the word "livingry," as the opposite of "weaponry"—which he called "killingry"—and popularized the term "spaceship earth." (He claimed to have invented "debunk," but probably did not.) Another one of his coinages was "ephemeralization," which meant, roughly speaking, "dematerialization." Fuller was a strong believer in the notion that "less is more," and not just in the aestheticized, Miesian sense of the phrase. He imagined that buildings would eventually be "ephemeralized" to such an extent that construction materials would be dispensed with altogether, and builders would instead rely on "electrical field and other utterly invisible environment controls."

Fuller's favorite neologism, "dymaxion," was concocted purely for public relations. When Marshall Field's displayed his model house, it wanted a catchy label, so it hired a consultant, who fashioned "dymaxion" out of bits of "dynamic," "maximum," and "ion." Fuller was so taken with the word, which had no known meaning, that he adopted it as a sort of brand name. The Dymaxion House led to the Dymaxion Vehicle, which led, in turn, to the Dymaxion Bathroom and the Dymaxion Deployment Unit, essentially a grain bin with windows. As a child, Fuller had assembled scrapbooks of letters and newspaper articles on subjects that interested him; when, later, he decided to keep a more systematic record of his life, including everything from his correspondence to his dry-cleaning bills, it became the Dymaxion Chronofile.

All the Dymaxion projects generated a great deal of hype, and that was clearly Fuller's desire. All of them also flopped. The first prototype of the Dymaxion Vehicle had been on the road for just three months when it crashed, near the entrance to the Chicago World's Fair; the driver was killed, and one of the passengers—a British aviation expert—was seriously injured. Eventually, it was revealed that another car was responsible for the accident, but only two more Dymaxion Vehicles were produced before production was halted, in 1934. Only thirteen models of the Dymaxion Bathroom—a single unit that came with a built-in tub, toilet, and sink—were constructed before the manufacturer pulled the plug on that project, in 1936. The Dymaxion Deployment Unit, which Fuller imagined being used as a mobile shelter, failed because after the United States entered the Second World War he could no longer obtain any steel. In 1945, Fuller attempted to mass-produce the Dymaxion House, entering into a joint effort with Beech Aircraft, which was based in Wichita. Two examples of the house were built before that project, too, collapsed. (The only surviving prototype, known as the Wichita House, looks like a cross between an onion dome and a flying saucer; it is now on display at the Henry Ford Museum, in Dearborn, Michigan.)

Following this string of disappointments, Fuller might have decided that his "experiment" had run its course. Instead, he kept right on going. Turning his attention to mathematics, he concluded that the Cartesian coördinate system had got things all wrong and invented his own system, which he called Synergetic Geometry. Synergetic Geometry was based on sixty-degree (rather than ninety-degree) angles, took the tetrahedron to be the basic building block of the universe, and avoided the use of pi, a number that Fuller found deeply distasteful. By 1948, Fuller's geometric investigations had led him to the idea of the geodesic dome—essentially, a series of struts that could support a covering skin. That summer, he was invited to teach at Black Mountain College, in North Carolina, where some of the other instructors included Josef Albers, Willem and Elaine de Kooning, John Cage, and Merce Cunningham. ("I remember thinking it's Bucky Fuller and his magic show," Cunningham would later recall of Fuller's arrival.) Toward the end of his stay, Fuller and a team of students assembled a trial dome out of Venetian-blind slats. Immediately upon being completed, the dome sagged and fell in on itself. (Some of the observers referred to it as a "flopahedron.") Fuller insisted that this outcome had been intentional—he was, he said, trying to determine the critical point at which the dome would collapse—but no one seems to have believed this. The following year, Anne Fuller sold thirty thousand dollars' worth of I.B.M. stock to finance Bucky's continuing research, and in 1950 he succeeded in erecting a dome fifty feet in diameter.

The geodesic dome is a prime example of "ephemeralization"; it can enclose more space with less material than virtually any other structure. The first commercial use of Fuller's design came in 1953, when the Ford Motor Company decided to cover the central courtyard of its Rotunda building, in Dearborn. The walls of the building, which had been erected for a temporary exhibit, were not strong enough to support a conventional dome. Fuller designed a geodesic dome of aluminum struts fitted with fibreglass panes. The structure spanned ninety-three feet, yet weighed just eight and a half tons. It received a tremendous amount of press, almost all of it positive, with the result that geodesic domes soon became popular for all sorts of purposes. They seemed to spring up "like toadstools after a rain," as one commentator put it.

The geodesic dome transformed Fuller from an eccentric outsider into an eccentric insider. He was hired by the Pentagon to design protective housing for radar equipment along the Distant Early Warning, or DEW, line; the structure became known as a radome. He also developed a system for erecting temporary domes at trade fairs all around the world. (Nikita Khrushchev supposedly became so enamored of one such dome, built for a fair in Moscow, that he insisted that "Buckingham Fuller" come to Russia "and teach our engineers.") Fuller was offered an appointment at Southern Illinois University, in Carbondale, and he had a dome-home built near campus for himself and Anne. In 1965, he was commissioned by the United States Information Agency to design the U.S. Pavilion for the Montreal Expo. Though the exhibit inside was criticized as uninspiring, Fuller's dome, which looked as if it were about to float free of the earth, was a hit.

As the fame of the dome—and domes themselves—spread, Fuller was in near-constant demand as a speaker. "I travel between Southern and Northern hemispheres and around the world so frequently that I no longer have any so-called normal winter and summer, nor normal night and day," he wrote in "Operating Manual for Spaceship Earth." "I wear three watches to tell me what time it is." Castro-like, Fuller could lecture for ten hours at a stretch. (A friend of mine who took an architecture course from Fuller at Yale recalls that classes lasted from nine o'clock in the morning until five in the evening, and that Fuller talked basically the entire time.) Audiences were enraptured and also, it seems, mystified. "It was great! What did he say?" became the standard joke. The first "Whole Earth Catalog," which was dedicated to Fuller, noted that his language "makes demands on your head like suddenly discovering an extra engine in your car."

In "Bucky," a biography-cum-meditation, published in 1973, the critic Hugh Kenner observed, "One of the ways I could arrange this book would make Fuller's talk seem systematic. I could also make it look like a string of platitudes, or like a set of notions never entertained before, or like a delirium." On the one hand, Fuller insisted that all the world's problems—from hunger and illiteracy to war—could be solved by technology. "You may . . . want to ask me how we are going to resolve the ever-accelerating dangerous impasse of world-opposed politicians and ideological dogmas," he observed at one point. "I answer, it will be resolved by the computer." On the other hand, he rejected fundamental tenets of modern science, most notably evolution. "We arrived from elsewhere in Universe as complete human beings," he maintained. He further insisted that humans had spread not from Africa but from Polynesia, and that dolphins were descended from these early, seafaring earthlings.

Although he looked to nature as the exemplar of efficient

design, he was not terribly interested in the natural world, and mocked those who warned about problems like resource depletion and overpopulation. "When world realization of its unlimited wealth has been established there as yet will be room for the whole of humanity to stand indoors in greater New York City, with more room for each human than at an average cocktail party," he wrote. He envisioned cutting people off from the elements entirely by building domed cities, which, he claimed, would offer free climate control, winter and summer. "A two-mile-diameter dome has been calculated to cover Mid-Manhattan Island, spanning west to east at 42nd Street," he observed. "The cost saving in ten years would pay for the dome. Domed cities are going to be essential to the occupation of the Arctic and the Antarctic." As an alternative, he developed a plan for a tetrahedral city, which was intended to house a million people and float in Tokyo Bay.

He also envisioned what he called Cloud Nines, communities that would dwell in extremely lightweight spheres, covered in a polyethylene skin. As the sun warmed the air inside, Fuller claimed, the sphere and all the buildings within it would rise into the air, like a balloon. "Many thousands of passengers could be housed aboard one-mile-diameter and larger cloud structures," he wrote. In the late seventies, Fuller took up with Werner Erhard, the controversial founder of the equally controversial est movement, and the pair set off on a speaking tour across America. Fuller championed, and for many years adhered to, a dietary regimen that consisted exclusively of prunes, tea, steak, and Jell-O.

The Dymaxion Vehicle, the Dymaxion House, "comprehensive, anticipatory design," Synergetic Geometry, floating cities, Jell-O—what does it all add up to? In conjunction with the Whitney retrospective, the exhibition's two curators, K. Michael Hays and Dana Miller, have put together a book of essays, articles, and photographs—"Buckminster Fuller: Starting with the Universe." Several of the authors in the volume gamely, if inconclusively, grapple with Fuller's legacy. Antoine Picon, a professor of architecture at Harvard, notes that the detail with which Fuller's life was recorded—the Dymaxion Chronofile eventually grew to more than two hundred thousand pages—has had the paradoxical effect of obscuring its significance. Elizabeth A. T. Smith, the chief curator at the Museum of Contemporary Art in Chicago, writes that Fuller's influence on "creative practice" has been "more wide-ranging than previously thought," but goes on to acknowledge that this influence is "difficult to pinpoint or define with certainty." In their introduction, Hays and Miller maintain that Fuller helped "us see the perils and possibilities" of the twentieth century. They stress his "continuing relevance as an aid to history," though exactly what they mean by this seems purposefully unclear.

The fact that so few of Fuller's ideas were ever realized certainly makes it hard to argue for his importance as an inventor. Even his most successful creation, the geodesic dome, proved to be a dud. In 1994, Stewart Brand, the founding editor of the "Whole Earth Catalog" and an early, self-described dome "propagandist," called geodesics a "massive, total failure":

> Domes leaked, always. The angles between the facets could never be sealed successfully. If you gave up and tried to shingle the whole damn thing—dangerous process, ugly result—the nearly horizontal shingles on top still took in water. The inside was basically one big room, impossible to subdivide, with too much space wasted up high. The shape made it a whispering gallery that broadcast private sounds to everyone.

Among the domes that leaked were Fuller's own home, in Carbondale, and the structure atop the Ford Rotunda. (When workmen were sent to try to reseal the Rotunda's dome, they ended up burning down the entire building.)

Fuller's impact as a social theorist is equally ambiguous. He insisted that the future could be radically different from the past, that humanity was capable of finding solutions to the most intractable-seeming problems, and that the only thing standing in the way was the tendency to cling to old "piano tops." But Fuller was also deeply pessimistic about people's capacity for change, which was why, he said, he had become an inventor in the first place. "I made up my mind . . . that I would never try to reform man—that's much too difficult," he told an interviewer for this magazine in 1966. "What I would do was to try to modify the environment in such a way as to get man moving in preferred directions." Fuller's writings and speeches are filled with this sort of tension, or, if you prefer, contradiction. He was a material determinist who believed in radical autonomy, an individualist who extolled mass production, and an environmentalist who wanted to dome over the Arctic. In the end, Fuller's greatest accomplishment may consist not in any particular idea or artifact but in the whole unlikely experiment that was Guinea Pig B. Instead of destroying himself, Fuller listened to Universe. He spent the next fifty years in a headlong, ceaseless act of self-assertion, one that took so many forms that, twenty-five years after his death, we are still trying to sort it all out.

Dan Hill

The Street as Platform

The way the street feels may soon be defined by what cannot be seen with the naked eye.

Imagine a film of a normal street right now, a relatively busy crossroads at 9AM taken from a vantage point high above the street, looking down at an angle as if from a CCTV camera. We can see several buildings, a dozen cars, and quite a few people, pavements dotted with street furniture.

Freeze the frame, and scrub the film backwards and forwards a little, observing the physical activity on the street. What can't we see?

We can't see how the street is immersed in a twitching, pulsing cloud of data. This is over and above the well-established electromagnetic radiation, crackles of static, radio waves conveying radio and television broadcasts in digital and analogue forms, police voice traffic. This is a new kind of data, collective and individual, aggregated and discrete, open and closed, constantly logging impossibly detailed patterns of behaviour. The behaviour of the street.

Such data emerges from the feet of three friends, grimly jogging past, whose Nike shoes track the frequency and duration of every step, comparing against pre-set targets for each

/cityofsound.com/

individual runner. This is cross-referenced with playlist data emerging from their three iPods. Similar performance data is being captured in the engine control systems of a stationary BMW waiting at a traffic light, beaming information back to the BMW service centre associated with the car's owner.

The traffic light system itself is capturing and collating data about traffic and pedestrian flow, based on real-time patterns surrounding the light, and conveying the state of congestion in the neighbourhood to the traffic planning authority for that region, which alters the lights' behaviour accordingly. (That same traffic data is subsequently scraped by an information visualisation system that maps average travel times on to house price data, overlaid onto a collaboratively produced and open map of the city.)

In an adjacent newsagent's, the stock control system updates as a newspaper is purchased, with data about consumption emerging from the EFTPOS system used to purchase the paper, triggering transactions in the customer's bank account records.

Data emerges from the seven simultaneous phone conversations (with one call via Skype and six cellular phones) amongst the group of people waiting at the pedestrian crossing nearest the newsagent.

The recent browser histories of the two PCs with Internet access in a coffee shop across the road update sporadically with use, indicating both individual patterns of websites accessed and an aggregate pattern of data transfer throughout the day. At the counter of the coffee shop, a loyalty card is being swiped, updating records in their customer database. The flat above the shop is silently broadcasting data about the occupant's usage of his Sky+ Box, DAB radio with Internet connection, and Xbox Live console. His laptop noisily plays music, noiselessly accreting data to build a profile of the user's taste in music at the web-based service Last FM. However, this particular track has inaccurate or no metadata; it is not registered by Last FM and this in turn harms its latent sales prospects.

A police car whistles by, the policewoman in the passenger seat tapping into a feed of patterns of suspicious activity around the back of the newsagent on a proprietary police system accessed via her secured BlackBerry. A kid takes a picture of the police car blurring past with his digital camera, which automatically uses a satellite to stamp the image with location data via the GPS-enabled peripheral plugged into the camera's hot-shoe connection.

Across the road, a telecom engineer secures a wireless device to the telephone exchange unit on the pavement, which will intermittently broadcast its state back to base, indicating when repairs might be necessary.

Walking past, an anxious-looking punter abruptly halts as the local Ladbrokes triggers a Bluetooth-based MMS to his phone, having detected him nearby, and offers discounts on a flutter on the 3.30 at Newmarket (the Ladbrokes is constantly receiving updates on runners, riders and bets, linked to a national network aggregating information from local nodes at racecourses and bookies). The potential punter had earlier received a tip on said race from his chosen newspaper's daily sports bulletin, delivered via his mobile's newsfeed reader software.

As he wonders whether he could discreetly sidle into the bookies to place the bet he'd promised himself he wouldn't, the street-lamp above his head fades down as its sensors indicate the level of ambient daylight on the street is now quite sufficient. It switches to another mode whereby the solar panel above collects energy for the evening and delivers any potential excess back into the grid, briefly triggering a message indicating this change of state back to the public-private partnership that runs the lighting services in this borough, in turn commencing a transaction to price up the surplus electricity delivered to the grid.

The same increase in daylight causes a minor adjustment in four of the seven CCTV cameras dotted along the street, as they re-calibrate their exposure levels accordingly. The digital video accruing on the array of remote disk-drives at a faraway control centre is rendered slightly differently in response.

In an apartment over the bookies, the occupant switches on her kettle, causing the display on her Wattson device that monitors real-time electricity usage in the flat to jump upwards by a kilowatt, whilst triggering a corresponding jump in the sparklines displaying usage on the Holmes software that tracks that data over time, and which compares her consumption to four of her friend's houses in the same neighbourhood.

Three kids are playing an online game on their mobile phones, in which the physical street pattern around them is overlaid with renderings of the 19th century city. They scuttle down an alleyway behind a furniture showroom as the virtual presence of another player, actually situated in a town forty miles away but reincarnated here as a Sherlock Holmes–style detective (indicated on their map by an icon of a deerstalker and gently puffing pipe), strides past the overlaid imagined space. The three play a trio of master criminals, intent on unleashing a poisonous miasma upon the unsuspecting and unreal caricatures generated by the game.

Approaching the furniture showroom's delivery bay to the rear, the driver of an articulated lorry grinds down through his gears in frustration as he realises the road over the lights narrows to a point through which his cab will not fit—information not made clear by the satnav system propped on his dashboard. The RFID chips embedded into the packaging of the seven armchairs in his trailer register briefly on the showroom's stock control system, noting the delivery time and identity of the driver. When the armchairs are formally checked in later by the showroom's assistant with a barcode-scanner, the damage to one of the chairs is noted and sent back to base, automatically triggering the addition of a replacement armchair to the next lorry out to this town, while recalculating stock levels.

In the shoe-shop next door, a similar hand-held scanner, unknowingly damaged in a minor act of tomfoolery a day earlier, fails to register the barcode on a box of sneakers, resulting in a lost sale as the assistant is unable to process the transaction without said barcode. The would-be customer walks out in disgust, texting his wife in order to vent his furious frustration on someone. She sends a placating if deliberately patronising message back within a few seconds, which causes him to smile and respond with an "x" two seconds after that. In doing so, he exceeds his allocation of SMSs for the month, tipping over to the next tier in his payment plan and triggering a flag in a database somewhere in Slough.

Deciding to spend his money—that he unwittingly has less of than he did a few moments ago—on a book instead, he steps into the only local bookstore on the street, using the now more expensive data plan of his mobile phone service to retrieve aggregated reviews for the latest Andy McNab, which he half-reads whilst perusing the back cover of the book. Unfortunately the corresponding prices offered up by the review system are in US dollars, as the service is not localised and thus he can't compare prices. This is fortunate for the shop, however, and so during the resulting purchase of the book, the store's stock control system automatically orders a fresh batch of the now best-seller, whilst the Top 10 display on the counter tracks McNab's seemingly inexorable rise up the charts on a battered old LCD monitor.

Round the corner, the number of copies of the McNab book in the municipal library remains exactly the same. Instead, the large external LED display hoisted over the door at huge expense conveys the volume of ISBNs of books being swiped by librarians inside the building, in real-time. Part of an installation by students at the local art college, the most popular genres of books taken out, inferred from the aggregate of ISBNs and cross-referenced with Amazon, are displayed every five minutes via a collage of randomly-selected movie clips from YouTube that match broadly that same genre and keywords (filtered for decency and sensitivity by bespoke software which is itself receiving updates, detailing what is considered obscene at this point). Currently, a 2-second sequence of a close-up of Alec Guinness's nose and moustache from *The Bridge on the River Kwai* morphs into the bulging right arm of Sylvester Stallone in *Rambo,* cradling a stolen Soviet rocket launcher. The patterns of clip consumption at YouTube twitch accordingly.

Looking up at the display in fascination and bewilderment, an elderly lady stumbles over a pothole in the pavement. Helped back to her feet by a younger man, she decides to complain to the council about the pothole. The man suggests he can do that right now, from his iPod Touch and using the library's open public wifi, by registering the presence of a pothole at this point on the local problems database, Fix My Street. The old woman stares at him quizzically as it takes him fifty seconds to close the website he had been looking at on his mobile (Google Maps directions for "hairdressers near SW4," a phrase he'll shortly have to type in again, having neglected to bookmark it) and access fixmystreet.com. He spends the next few minutes indicating the presence of a pothole outside the library on Fix My Street (unaware of the postcode, he has to select one from a few possible matches on street name), before he moves on, satisfied with his civic good deed for the day. The elderly lady had long since shuffled off, muttering to herself. Although Fix My Street smartly forwards on all issues to the corresponding council, a beleaguered undertrained temp in the also underfunded "pavements team" is unaware of fixmystreet.com and unable to cope with the levels of complaint, and so the pothole claims five more victims over the next two weeks until someone rings up about it.

The LED display board can also sniff what is being accessed via the library's public wifi network, and displays fragments of the corresponding text and imagery. It switches briefly over to this mode, in order to denote that Fix My Street was being accessed, and displays some details of the transactions detailing the pothole issue. Before flicking back to the YouTube x ISBN installation, the display then conveys some information from the local council about a forthcoming street upgrade, blissfully unaware of the possible connection to be made between that and the pothole. Unfortunately, at that point, the pale sunlight hits the screen at such an angle that the two hurrying passers-by cannot read it anyway. The display then dissolves into a slow pan across Keira Knightly's delicately arched eyebrow from *Pirates of the Caribbean.*

In the swinging briefcase of one of the passers-by, an Amazon Kindle e-book reader briefly connects to the public library; having previously visited the library, the owner had registered the public wifi in her settings. It commences a rapid-fire series of handshakes with Amazon's systems, swapping personal details back and forth with user profile information, and thus beginning to download a new book by Ian McEwan to the device. Despite the wealth of metadata in this rich stream of data, the Kindle's closed system means that the library's databases, and LED display installation, cannot possibly be made aware of this literary transaction being conducted using its infrastructure. Either way, in seven seconds the Kindle user is out of range and the download automatically fizzles out, settling back to wait for the scent of open wireless.

Behind the library, a small 19th century cottage that's been on the market for a year now is being re-valued by estate agents.

This new figure, a few thousand pounds less than the previous one, is entered automatically via the estate agent's PDA and ripples through their internal databases and then external facing systems. But it doesn't trigger any change in three other proprietary databases listing average house prices in the neighbourhood until three weeks later. This house's change in price subtly affects the average for the area, which is later recombined into the aforementioned map that compares with commuter times for the borough.

An employee of the local water company knocks on a door up the street, calling in order to take a reading from the house's meter. She uses a bespoke application on her mobile phone, which should indicate the location of the meter on the property. In this case, it doesn't, so she has to ask.

Five TomTom satnav systems in five of the twelve cars on the street suddenly crash for reasons unknown, causing an instantaneous reboot and login sequence over the course of twenty seconds. One driver notices.

The four other drivers are slightly distracted by the glow of a giant TV screen, installed and operated by the council but paid for through corporate sponsorship, which glowers over the end of a pedestrianised-shopping mall at the end of the street over the lights. It's broadcasting the Australian Open tennis, which is being played live in Melbourne. A homeless person is sleeping underneath the screen, soaking up some of its transmitted warmth. An on-street information kiosk stands beside the screen, offering a scrollable map of the local area and directory of local businesses. It's rarely used, as the directory of businesses was always incomplete and intermittently updated, its data now rusty and eroded by time. Plus, maps are now available on most people's mobile phones. Still, the printer installed in its base occasionally emits a money-off coupon for some of the local businesses.

Under the pavement on one side of the street, a buried sensor records the fact that some fibre-optic cables are now transmitting data 10% less effectively than when they were installed. A rat ascends from an accidentally uncovered grille under the library's down-pipe nearby, its whiskers containing trace elements of plastic cladding.

A blogger posts an entry on her weblog regarding some new graffiti on the library's rear, uploading the image via her mobile phone, thanks to her blog platform's relationship with Flickr, a popular photo sharing site. She adds a cursory description of the stencilled representation of the Mayor's face superimposed onto a £50 note instead of the Queen's. Shortly afterwards, she receives an SMS from the service Twitter, indicating that two of her friends are heading for a café up the street, and she decides to intercept them to discuss her find, sending back the URL of her post and the time of her imminent arrival. Her phone's Google Maps application triangulates her position to within a few hundred metres using the mobile cell that encompasses the street to convey a quicker route to the café. Unfortunately, none of their systems convey that the café is newly closed for redecoration.

Working from home in his small house backing onto the old cottage, a lawyer files his case notes via the password-protected intranet his company operates. His wifi network is encrypted to prevent leakage of such confidential data. He then closes his network connection, switching instead to his neighbour's wifi network (which has been left purposefully open in the interests of creating a cohesive civic layer of wireless coverage on their street) in order to watch the highlights of his football team's two-nil victory the night before. In this way, his own remarkably cheap wireless network data plan never goes beyond its monthly cap. This parasitic wireless activity is only curtailed months later, when the previously benevolent neighbour uses some free sniffer software she downloaded to detect the presence of the wifi router that's responsible for the majority of the data usage in the street.

A local off-license has an old monitor in the window that cycles through a series of crude screen-grabs of faces of shoplifters of local stores, derived from the various CCTV systems owned by a local association of shopkeepers. Unfortunately, the face of the purchaser of the Andy McNab book is mistakenly added to the system three weeks later.

(Coincidentally, in a meeting being conducted several miles away, a project team working on council tax systems briefly considers whether a system of localised screens displaying which houses in the street had not paid their council tax yet, updated wirelessly, would be ethically sound.)

Waiting at the lights, someone pays their council tax by mobile phone, triggering an Internet-based bank transfer via SMS. Across the road, a car belonging to a city-wide car-sharing network patiently waits to be activated by a swipe of a member's RFID card. It transmits its location and status back to the car-sharing network's database every few minutes.

Also in a prime position by the lights, a café is briefly office to two business-people having an informal meeting. Although the café's wireless network is closed, their usage charges are paid by the company they work for, and they barely notice the cost. The company credit card details are retrieved automatically over a secure transaction. Though it has poorer muffins, the café opens 90 minutes earlier than the library.

A series of small high-resolution displays, hanging under each traffic light and angled towards stationary drivers, alternately communicates the number of accidents that have occurred around these lights in the last year, and then the current speed limit, which can be calibrated to an optimum level for the current traffic conditions in the borough. The traffic lights also house the city's congestion charging system's cameras, logging the number-plates of cars passing through its network of inner-city streets.

A wireless sensor network, carefully and discreetly embedded in the trunks of trees lining one side of the street, silently monitors the overall health of the limes and planes, collating data and waiting patiently for the council's tree surgeon to inspect the arboreal vital signs.

At the end of the line of trees, a new bench has been installed on the street. At either side of the bench, there are two standard electrical power-points freely available for citizens to recharge their phone or laptop. A small LED winks to indicate this, alongside a standardised explanatory icon drawn up by the department also responsible for the highways' signage systems. The power running to the bench is carried via flexible cables that can twist and stretch around the growing roots of the nearest trees. The bench also carries a WIMAX transmitter as part of a research project led by the local university. As a result of this, the bench appears as a key node on several GIS.

A cab drives through the traffic lights as they switch to green and it quickly signals to turn left, looking to nose back on itself as the presence of a fare is indicated at a nearby hairdressers, via the in-taxi control system. A faraway voice crackles over the intercom a few seconds later attempting to verify that the driver is en-route. The driver clarifies she is en-route but that she'll take a few minutes more than usual as her satnav system indicates high traffic levels across the three normal routes taken.

At another building on the street, a new four-storey commercial office block inhabited by five different companies, the building information modelling systems, left running after construction, convey real-time performance data on the building's heating, plumbing, lighting and electrical systems back to the facilities management database operated by the company responsible for running and servicing the building. It also triggers entries in the database of the engineering firms who designed and built the office block, and are running post-occupancy evaluations on the building in order to learn from its performance once inhabited.

In turn, and using this feed, the city council's monitoring systems note the aggregated energy usage for the commercial buildings on the street, constantly shuffling its league table of energy-efficient neighbourhoods. The current score for the street, neighbourhood and city is displayed outside the nearby library, on a trio of vertical axis wind turbines with LEDs embedded in their blades.

A prototype of a similar monitoring system, but embedded in the bus-stop opposite the library, records the performance of the lights, travel information displays, large plasma-screen advertising display, and the chilled-beam cooling system newly installed for comfort. The travel information displays themselves receive updates in real-time via a slice of radio spectrum allocated to such data, indicating the proximity of the next five buses. This same system also conveys the latest information on the whereabouts of the no. 73 in particular, in the form of an SMS to a prospective passenger who has selected this as her "favourite bus" via the transport company's website. Around the corner, she breaks into a trot accordingly.

The plasma display is currently running an advert for the local radio station's breakfast show (displaying a live stream of SMS messages sent to the show, filtered for obscenity and likelihood of libel). As the slightly out-of-breath imminent passenger arrives within range of its Bluetooth-based transceiver, it cross-fades to a display from the city's premiere modern art gallery, with which she has registered her mobile phone as a preferred mode of communication and whose systems are quickly cross-referenced for her attendance record for the last few years, and thus the bus-stop informs her of a new exhibition about to start.

This she doesn't notice at all, but one person in the loosely-defined queue around the bus stop does, and scribbles the details on his hand. Four seconds later, the display recognises another mobile phone with an open Bluetooth connection and an active account within the agglomeration of companies that have registered their databases with this advertising service, and shifts its display accordingly. The call-and-response between the queue and the screen continues until the bus finally pulls in and the screen's transient audience dissipates. It settles back to a carousel of generic advertising messages and local information tailored to that street and its surrounds.

As the bus departs, the new passengers on-board swipe their RFID-based integrated transport system ID cards, updating mass transit databases with every possible aspect that can be gleaned from this simple activity (time of day, location, frequency of use, favourite entry points etc.). Using simple sensors, the now-empty seat in the bus-stop registers that it is indeed now empty, and wirelessly logs this fact with a database that monitors the usage and state of street furniture in the neighbourhood. Powered by solar panels on top of the bus-stop, it creates a pulsing ambient glow.

Across the road, another billboard displays the number of reported burglaries and bag-snatches in the neighbourhood in the last three months: live data direct from the police force systems. This causes several passers-by to feel a touch more anxious than they did a moment ago. Had they walked past a moment before, the billboard would have been displaying information on a forthcoming community sports day at the local park. One of the passers-by would have recognised their son in the video of last year's winners, running in slow motion under the crisp typography. A moment before and the passers-by would have been subjected to a tortuous promo for a Portuguese avant-garde play currently running at the local theatre, within which a QR code displayed in the top-right hand corner could've been read with a mobile phone's IR reader, delivering the website for the theatre to the phone's browser.

Of the two bars, two pubs and three cafés on the street, only one has recently checked that the location and description data overlaid on Google Maps is present and correct, and is thus fortunate to receive the custom of two hungry Hungarian tourists for a full English breakfast with all the trimmings.

Twenty metres below the ground, a tube train scurries under the crossroads, outrunning the halo of data that details its location and speed from the engine control systems, while CCTV conveys images of the carriage directly underneath. The carriage contains forty-four mobile phones seeking a signal, some with Bluetooth headphone sets; ten BlackBerries and four other PDAs likewise; thirteen mp3 players of varying brands, a couple also with Bluetooth headphones; seven sleeping laptops.

Directly overhead, ten thousand metres up, the distant roar of a commercial airliner's Rolls-Royce engines, beaming their performance data back to engineers via satellite in real-time . . .

And press play . . .

The rest of this essay can be accessed at www.cityofsound.com.

Sharon Weinberger

Can You Spot the Chinese Nuclear Sub?

Widely available satellite imagery is making governments around the world awfully nervous.

In a generic-looking glass and concrete office building just a few miles from Washington Dulles International Airport, an independent company is—with the full blessing of the government—helping to peel away the last earthly vestiges of cold war secrecy. Unlike Beltway defense companies, where security often starts at the ground floor with guards and gates, GeoEye, one of two U.S. companies selling commercial satellite imagery (the other is DigitalGlobe), is notably open-door until you reach the fourth-floor offices, where a friendly secretary asks, apologetically, whether the visitor is a U.S. citizen.

Inside the company's headquarters, GeoEye vice president Mark Brender provides a tour of operations. Young engineers in a NASA-style mission control room follow a GeoEye satellite, which is at that moment hurtling around Earth some 423 miles above the ground. Across the hall, GeoEye employees field calls from customers ranging from government agencies to insurance companies. CNN plays continuously on the TV so employees can be alerted to a crisis—a flood, a North Korean nuclear test, a border skirmish—and quickly send orders to the satellite to capture pictures of a specific site.

/Discover Magazine/

For a few thousand dollars, pretty much any American can buy up-to-the-moment satellite images of Iran's nuclear sites, CIA headquarters, even the top secret Air Force testing site, Area 51, in Nevada. Short on cash? If you don't mind older images, you can view these same sites for free on platforms like Google Earth, the ever-expanding Google service that uses 3-D visualization software to zoom in on different parts of the globe and deliver images to any PC hooked up to the World Wide Web.

This kind of imagery was, for much of the 20th century, part of the eyes-only world of intelligence; from the design of a nuclear submarine to the movement of Israeli troops, one needed high-level clearance for a glimpse. Not anymore. Today, with the advent of civilian satellites here and abroad, we have opened wide the window on places and events that, not so long ago, only spies could see.

Governments around the world have often reacted with outrage to the new age of transparency. And the Pentagon, while hardly thrilled, has had to adjust.

Without a doubt, "the technologies make clandestine operations more difficult to hide and raise the specter of enemies' having more capability to target critical nodes," says Theresa Hitchens, a space policy analyst and director of the Washington-based Center for Defense Information.

Yet the expanding field of vision also augurs a unique counterbalance to government excesses of the past. "It improves international security by allowing nongovernmental actors to monitor the actions of nations," says Hitchens, who points to Amnesty International; that group has launched eyesondarfur.org, a Web site that broadcasts satellite imagery of 12 vulnerable villages in war-torn western Sudan, empowering the public to document atrocities and track the movement of refugees and troops.

For the good and the bad, the lightless world has been illuminated, and the age of transparency is upon us. "The technologies are out of the bag, and governments will not be able to control access everywhere," Hitchens says. Keith Clarke, a University of California at Santa Barbara geography professor deeply involved in analyzing declassified satellite imagery, agrees. "It's a lost cause," he says of government efforts to stanch the flow of sensitive images. "There's very little that can be done to prevent it now."

DAWN OF AN ERA

The imagery we take for granted across our computer screens traces its origins to 1959 and the cold war, when the United States launched a top secret satellite called Corona. Corona's pair of constantly rotating cameras captured images on film cassettes that were parachuted back to Earth. Since being declassified in 1995, many of Corona's details have come to light. The earliest Corona images captured objects at a ground resolution of up to 8 meters (about 25 feet), meaning objects on the ground had to be at least that length to be discernible. Although grainy by current standards, the images were still a boon to U.S. intelligence, allowing the United States to look deep into Soviet territory, one time capturing images of a military base that helped dispel the notion of a "missile gap" between the United States and the then U.S.S.R.

Corona was just the start. With the development of light-sensitive solid-state chips in the 1970s, the spy community turned to reconnaissance satellites that could dispense with parachutes and beam images to Earth. These so-called electro-optical satellites produced images at resolutions better than 1 meter (about 3 feet)—detailed enough to show individual cars and revealing such secrets as a new Soviet aircraft carrier under construction at a Black Sea shipyard.

The civilian world was also getting into the act, and its most notable early achievement was SPOT, launched by the French in 1986. With 10-meter ground resolution, SPOT was hardly a match for military counterparts, but it was certainly good enough for the press to embrace its images and break news. In one instance, journalists used the images to document the construction of the Soviet Krasnoyarsk radar, an alleged violation of the antiballistic missile treaty, explains nuclear physicist Peter D. Zimmerman, former science adviser for arms control in the U.S. State Department.

By the 1990s foreign countries including France and Russia were offering satellite imagery for sale at increasingly crisp resolutions of up to five meters. A critical turning point, according to Zimmerman, was the Persian Gulf war, when Mark Brender, then a journalist at ABC News, obtained high-resolution satellite images of Kuwait from Russia, and another journalist, Jean Heller of the St. Petersburg Times, made those images public. The revelations were critical: The U.S. government had claimed that Iraq had around 250,000 troops in Kuwait, but the evidence revealed just a fraction of that number.

Some hoped to rein in the imagery and all the associated leaks—but it was a hopeless cause. Today, supported and even encouraged by a U.S. government hoping to sharpen the national edge, companies can sell imagery to private customers here and abroad, though not to what are dubbed "bad actors," such as Iran. Companies in the United States also cannot sell satellite imagery with a resolution better than about 1.5 feet (0.5 meter), except to the U.S. government. Resolution better than half a meter is "seen as having significant national security value," explains Kay Weston, who manages the licensing process at the National Oceanic and Atmospheric Administration. "That's the tipping point."

Maintaining even minimal security may be hard because, in the last couple of years, the bad actor clause, the resolution limit, and other protections have been challenged by easy access to Internet services like Google Earth. This open-access technology combines satellite images with even better pictures taken from planes and on the ground. The irony of Google Earth—something not lost on national security agencies—is that the technology itself is a product of government support. In 2003 the National Imagery and Mapping Agency (now known as the National Geospatial-Intelligence Agency) began investing in a new company called Keyhole, which melded satellite imagery and photographs taken from aircraft into a three-dimensional tool for nuanced mapping of all kinds of terrain, including military areas and city streets.

Google bought Keyhole in 2004 and rolled out Keyhole's software as Google Earth, opening the door to a new set of national security questions. Especially for amateurs, the power of the tool is extraordinary. Once Internet users download the free software, Google Earth enables them to swoop in on their own home simply by entering a street address or GPS coordinates, creating the visual sensation of flying from outer space to a specific point on Earth in a split second. Once you reach the designated site, you can zoom in and out for more or less detail or click on screen buttons that allow you to explore the surrounding area.

Not only is the tool simple to use, but it is available to anyone, anywhere in the world. While U.S. companies must observe government restrictions on selling images to certain countries and individuals, Google Earth just puts the images online, enabling unrestricted access. How then can bad actors be stopped from viewing imagery that they cannot legally buy? According to Weston, some protection accrues from the simple profit motive. "Companies don't want to give away current imagery for free," she says. The satellite imagery on Google Earth ranges from a few months to three years in age.

Yet old satellite pictures become more powerful when combined with some of the other free offerings from Google Earth, such as high-resolution aerial photographs (from planes or helicopters). Sometimes such images are freely available directly from the government, and other times they are purchased from private companies that take the photos for a fee. To expand this capability, in 2007 Google bought Image-America, a private company that provides aerial shots. Indeed, while the U.S. government prohibits sale of satellite imagery with ground resolution better than a half-meter, no such rule applies to images from nonsatellite sources. So while a satellite company may be forced to "fuzz up" an image of, say, CIA headquarters in McLean, Virginia, to meet the half-meter standard, that same picture is available in even sharper focus from aerial photographs on Google Earth.

"I worry about the self-delusion factor," Hitchens says. "It's almost like willing ourselves to believe no one can see what we're doing. That's patently not true."

THE BREACHES

With the proliferation of online tools, it is hardly surprising that state secrets have been exposed. In a highly public—and embarrassing—incident, a blogger in July 2007 noticed a photo showing a U.S. Navy nuclear submarine in Bangor, Washington, posted on Microsoft-owned Virtual Earth, a Google Earth rival. The problem was not the sub per se, but the sub's propeller, which was supposed to be covered with a shroud when out of water to protect its top secret design. The propeller was left exposed, and the image was posted online.

"They protect those propellers like it's the body and blood of Christ, the host," says John Pike, who runs Globalsecurity.org and was one of the first private analysts to buy commercial satellite imagery. "It had been 30 years since anyone had seen an American nuclear submarine propeller."

Nor is the U.S. government the only one to get caught up in the expanding reach of freely available imagery. Israel, in particular, has fought to control high-resolution imagery of its territory. More than a decade ago, the U.S. Congress passed the Kyl-Bingaman amendment, which prohibits U.S. companies from selling commercial satellite imagery of Israel that is better than similar imagery from other commercial sources; in practice this has meant that imagery of Israel is degraded to two meters, a restriction not imposed on any images of the United States.

In another notable case, a U.S. nuclear expert last year pointed to satellite images that appeared online showing China's new nuclear ballistic missile submarine. "Both China and India have raised their hackles about imaging satellites and tried to push back," Hitchens says. Calls by these countries to restrict satellite imagery, if heeded, would be a "scary" precedent, she says.

Some incidents are less about national security than about embarrassment: Bahrain's royal family briefly fought to block its citizens from viewing Google Earth images of its luxury housing, for fear of personal exposure.

But perhaps the most telling example of how the rapid expansion of mapping technology has caught the Pentagon off guard is the case of Google Street View, a feature Google introduced in 2007 that includes street-level photos providing 360-degree views, allowing users to weave their way through a city. Google collects the pictures for Street View by sending out cars and vans equipped with a special camera that captures the distinctive panoramic images. While Google Street

View raised privacy concerns for individuals captured in the images, the novel addition to Google was not something that immediately raised red flags for national security.

Then, in late February, just weeks after returning from his second deployment to Iraq, Colonel James Brown, who heads force protection and mission assurance at U.S. Northern Command in Colorado Springs, received an unexpected message: Was he aware that something called Google Street View had mapped out all the roads at Fort Sam Houston, an Army base in Texas?

Brown wasn't familiar with Google Street View, so he went online to take a look for himself. There it was: panoramic views of the entire base; you could even zoom in and out on security fences and guards. He recalled thinking to himself, "Oh, my God, this is the best preoperational surveillance tool I've ever seen in my life."

In other words, it was ideal for terrorists planning an attack on the base. "I was aghast," Brown recalls. "We didn't have any idea that this existed." Brown decided he needed to figure out what sort of online images might prove a threat to national security.

At U.S. Northern Command, which was established after 9/11 to defend the homeland, Brown put the critical infrastructure protection team to work, setting them loose on the Internet. "What was funny," he says, "is that the first night, the answer came back: This is the list of DOD [Department of Defense] facilities that are mapped, and we thought, 'Oh, boy, this is bad.'" It turned out to be a false alarm; a number of the flagged facilities were military bases that had been decommissioned, so their mapping was not a threat. Brown's group also looked at photographs of entrances to bases from the outside but decided to let those cases go. "We didn't want to get involved in trying to pull the wings off photography flies off base," he says.

In the case of Fort Sam Houston, the panoramic street

views ended up being something of a fluke; Google mappers had mistakenly been allowed on base. (Brown says that at the time, base security had seen the street mapping program as innocuous.) Within days of a review, however, U.S. Northern Command issued an order to U.S. bases and facilities instructing them not to allow this in the future. The command also contacted Google, which quickly removed the Street View images of Fort Sam Houston.

THE FUTURE OF I SPY

The goal of national security entities has long been to know everything, everywhere, at all times—a concept dubbed "universal situation awareness" by one former Air Force official. Toward that end, technology is advancing at a rapid pace.

State-of-the-art technology still belongs to the secret world of the intelligence community, and satellite experts familiar with classified technology are hesitant to speculate about precise capabilities. "I don't believe that the optimal performance of systems we have up there right now is public information," says Sidney Drell, professor emeritus at the Stanford Linear Accelerator Center, who is credited as one of the fathers of satellite reconnaissance. Outside analysts like Steven Aftergood of the Federation of American Scientists say that a resolution of 10 centimeters—allowing us to see a softball or the fine details of a car from space—is likely the best available resolution using visible light. This limit is imposed by the atmosphere, which diffracts visible light beyond that point.

To gather other kinds of data, meanwhile, some technologies focus on alternate regions of the electromagnetic spectrum, for instance, radar. Radar satellites come equipped with an antenna that sends radio waves to Earth; after hitting the planet's surface, the signal is reflected and scattered back toward a detector that generates an image without the need for visible light.

"You can see day and night and through most weather," says David Glackin, a remote sensing scientist with the Aerospace Corporation, a federally funded nonprofit research center.

Several nations, including the United States, have already launched radar-imaging satellites, but efforts abound to push the envelope further. One ambitious Air Force project involves a constellation of space-based radar satellites that would surround the planet, providing full coverage with the kind of granular detail thus far available only through aircraft or vehicles on the ground. The idea is to build a reconnaissance net that would target troops and even individuals from second to second, across every square foot of the globe. Gathering so much information has unique requirements, including cameras with huge apertures. The project "got a lot of criticism from academics who showed that the approach wasn't going to work unless the face of the detector in space was as long as a football field," says Philip Coyle, the Pentagon's former top technology tester and a senior adviser to the Center for Defense Information. "Something that huge makes a very big target for an enemy."

In the meantime, the image resolution of commercial satellites continues to improve. GeoEye's IKONOS, currently its premier satellite, provides resolution at better than one meter, but the company plans this August to launch GeoEye-1, a polar orbiting satellite equipped with a camera that can capture up to 700,000 square kilometers of black-and-white and 350,000 square kilometers of color imagery a day. The satellite, equipped with GPS, will be able to swivel and point its camera with a ground resolution of 41 centimeters.

"From 423 miles in space, we'll be able to see an object the size of home plate on a baseball diamond," says Brender. "GeoEye-1 will have the best accuracy and resolution of any commercial imaging satellite in the world when it goes up," says Bill Schuster, GeoEye's CEO.

For nongovernment customers, however, GeoEye-1 will still have to degrade that high-quality imagery to half a meter. The main barrier for commercial companies like GeoEye and DigitalGlobe, which supplies satellite imagery to Google, is not technology but economics and policy; there's no point taking sharper pictures if they aren't allowed to sell them widely. Some of the best future images, meanwhile, may come from foreign competitors focused on increasing satellite resolution themselves. As Glackin points out, in 1980 only five countries were operating imaging satellites. "Today there are 31," he says.

TRANSPARENT PLANET

The advent of Google Earth and related services raises a basic question: Have the images created a security risk? The answer, according to a number of experts, is yes, but there's not much that can be done about it. It's not clear that the U.S. government has the authority to remove imagery already in the public domain, according to Aftergood. And the availability of imagery from foreign sources may make some cases moot. "I think the answer probably is case specific, and only exceptional cases could qualify for legal or official intervention," he suggests.

Attempts to control satellite imagery are "a losing proposition," agrees Jeffrey Richelson, a senior fellow at the National Security Archive and a longtime expert on satellite reconnaissance. "These capabilities keep improving, and the sources of imagery keep expanding; look at countries that have launched high-resolution commercial imagery satellites over the last years. As that number goes up, it's difficult to prevent people who want to get images of particular locations from getting them."

And it is no longer just overhead imagery that is increasingly available. What makes the case of Google Street View and the mapping of Fort Sam Houston unique is precisely that it involved not overhead satellite imagery but rather cameras on the ground, raising more questions about Pentagon control as technology develops faster than the government's ability—or authority—to regulate it. Colonel Brown, for his part, acknowledges that tension. "What we're wrestling with is the fact that our technology is developing at such a rapid rate, it is crossing over boundaries we took for granted," he says.

Some countries may still be battling the onslaught of available imagery, but here in the United States, the Pentagon has realized the futility of de-Googling Earth: Both U.S. Northern Command and the National Geospatial-Intelligence Agency say they don't actively trawl the Internet looking for images of sensitive U.S. sites, although commercial satellite service could be interrupted if profound risk were perceived. When asked whether GeoEye is prohibited from selling imagery of any sensitive areas, Brender said no. As proof, he e-mailed, in midconversation, a detailed picture of Area 51, showing new construction at the Air Force's top secret testing facility in Nevada, including new hangars and a lengthened runway.

In the GeoEye conference room in Virginia, Brender flips through slides of imagery depicting the world's most mapped areas, including conflict hot spots, borders, and cities that are home to millions. These are the places people want to see. Could someone obtain such imagery as part of a terrorist plot? "We are not aware of one instance where our imagery was used for the greater evil," Brender says.

But even if images were misused, it would not change the trend. We've already entered the era of extreme transparency, and given the widespread technology, our clarity of vision can only expand. With the once-shrouded world bathed in so much light, Brender says, it would in the long run be "fruitless to try to hamper how we look at Earth from nonsovereign space."

It's not just governments that have problems hiding from the modern panopticon of satellites, aircraft, and ground-based cameras: Private citizens, too, are exposed. "You don't have exclusive rights to your property from the street or from above," says UCSB geography professor Keith Clarke. "Anybody who wants can take a photograph of your house. Many groups are doing exactly that."

So is there a way to hide from the infinite reach of Google Earth? You could choose to live someplace that people don't usually want to see in high resolution: a rural or sparsely populated area far from urban centers, international borders, and major points of interest. You could live underground, in a subway tunnel or a cave. Even in the suburbs, you could live under a canopy of trees. To avoid having video images of your house popping up on Google Street View, meanwhile, the best protection could be surrounding your property with a very high fence.

Your house is not the only thing that could show up online: Street View captures full-face shots of private citizens who happen to amble into the field of vision of its voracious video cameras. So to keep your face out of the database, stay alert. If you happen to see the telltale mounted videocam of Google Street View (it can be atop any kind of vehicle from a subcompact to a minivan), shield yourself with a sweatshirt hood, a newspaper, or even your hands.

There is one residence on the planet famously obscured by Google Earth: the Naval Observatory, official home of the vice president of the United States. Yet even Dick Cheney can't escape the long arm of the Internet—a clear image of the Naval Observatory is available on Yahoo.

Kevin Kelly

Becoming Screen Literate

Screens everywhere require screen fluency.

Everywhere we look, we see screens. The other day I watched clips from a movie as I pumped gas into my car. The other night I saw a movie on the backseat of a plane. We will watch anywhere. Screens playing video pop up in the most unexpected places—like A.T.M. machines and supermarket checkout lines and tiny phones; some movie fans watch entire films in between calls. These ever-present screens have created an audience for very short moving pictures, as brief as three minutes, while cheap digital creation tools have empowered a new generation of filmmakers, who are rapidly filling up those screens. We are headed toward screen ubiquity.

When technology shifts, it bends the culture. Once, long ago, culture revolved around the spoken word. The oral skills of memorization, recitation and rhetoric instilled in societies a reverence for the past, the ambiguous, the ornate and the subjective. Then, about 500 years ago, orality was overthrown by technology. Gutenberg's invention of metallic movable type elevated writing into a central position in the culture. By the means of cheap and perfect copies, text became the engine of change and the foundation of stability. From printing came journalism, science and the mathematics of libraries and law.

The distribution-and-display device that we call printing instilled in society a reverence for precision (of black ink on white paper), an appreciation for linear logic (in a sentence), a passion for objectivity (of printed fact) and an allegiance to authority (via authors), whose truth was as fixed and final as a book. In the West, we became people of the book.

Now invention is again overthrowing the dominant media. A new distribution-and-display technology is nudging the book aside and catapulting images, and especially moving images, to the center of the culture. We are becoming people of the screen. The fluid and fleeting symbols on a screen pull us away from the classical notions of monumental authors and authority. On the screen, the subjective again trumps the objective. The past is a rush of data streams cut and rearranged into a new mashup, while truth is something you assemble yourself on your own screen as you jump from link to link. We are now in the middle of a second Gutenberg shift—from book fluency to screen fluency, from literacy to visuality.

The overthrow of the book would have happened long ago but for the great user asymmetry inherent in all media. It is easier to read a book than to write one; easier to listen to a song than to compose one; easier to attend a play than to produce one. But movies in particular suffer from this user asymmetry. The intensely collaborative work needed to coddle chemically treated film and paste together its strips into movies meant that it was vastly easier to watch a movie than to make one. A Hollywood blockbuster can take a million person-hours to produce and only two hours to consume. But now, cheap and universal tools of creation (megapixel phone cameras, Photoshop, iMovie) are quickly reducing the effort needed to create moving images. To the utter bafflement of the experts who confidently claimed that viewers would never rise from their reclining passivity, tens of millions of people have in recent years spent uncountable hours making movies of their own design. Having a ready and reachable audience of potential millions helps, as does the choice of multiple modes in which to create. Because of new consumer gadgets, community training, peer encouragement and fiendishly clever software, the ease of making video now approaches the ease of writing.

This is not how Hollywood makes films, of course. A blockbuster film is a gigantic creature custom-built by hand. Like a Siberian tiger, it demands our attention—but it is also very rare. In 2007, 600 feature films were released in the United States, or about 1,200 hours of moving images. As a percentage of the hundreds of millions of hours of moving images produced annually today, 1,200 hours is tiny. It is a rounding error.

We tend to think the tiger represents the animal kingdom, but in truth, a grasshopper is a truer statistical example of an animal. The handcrafted Hollywood film won't go away, but if we want to see the future of motion pictures, we need to study the swarming food chain below—YouTube, indie films, TV serials and insect-scale lip-sync mashups—and not just the tiny apex of tigers. The bottom is where the action is, and where screen literacy originates.

An emerging set of cheap tools is now making it easy to create digital video. There were more than 10 billion views of video on YouTube in September. The most popular videos were watched as many times as any blockbuster movie. Many are mashups of existing video material. Most vernacular video makers start with the tools of Movie Maker or iMovie, or with Web-based video editing software like Jumpcut. They take soundtracks found online, or recorded in their bedrooms, cut and reorder scenes, enter text and then layer in a new story or novel point of view. Remixing commercials is rampant. A typical creation might artfully combine the audio of a Budweiser "Wassup" commercial with visuals from *The Simpsons* (or the

Teletubbies or *Lord of the Rings*). Recutting movie trailers allows unknown auteurs to turn a comedy into a horror flick, or vice versa.

Rewriting video can even become a kind of collective sport. Hundreds of thousands of passionate anime fans around the world (meeting online, of course) remix Japanese animated cartoons. They clip the cartoons into tiny pieces, some only a few frames long, then rearrange them with video editing software and give them new soundtracks and music, often with English dialogue. This probably involves far more work than was required to edit the original cartoon but far less work than editing a clip a decade ago. The new videos, called Anime Music Videos, tell completely new stories. The real achievement in this subculture is to win the Iron Editor challenge. Just as in the TV cookoff contest *Iron Chef,* the Iron Editor must remix videos in real time in front of an audience while competing with other editors to demonstrate superior visual literacy. The best editors can remix video as fast as you might type.

In fact, the habits of the mashup are borrowed from textual literacy. You cut and paste words on a page. You quote verbatim from an expert. You paraphrase a lovely expression. You add a layer of detail found elsewhere. You borrow the structure from one work to use as your own. You move frames around as if they were phrases.

Digital technology gives the professional a new language as well. An image stored on a memory disc instead of celluloid film has a plasticity that allows it to be manipulated as if the picture were words rather than a photo. Hollywood mavericks like George Lucas have embraced digital technology and pioneered a more fluent way of filmmaking. In his *Star Wars* films, Lucas devised a method of moviemaking that has more in common with the way books and paintings are made than with traditional cinematography.

In classic cinematography, a film is planned out in scenes; the scenes are filmed (usually more than once); and from a surfeit of these captured scenes, a movie is assembled. Sometimes a director must go back for "pickup" shots if the final story cannot be told with the available film. With the new screen fluency enabled by digital technology, however, a movie scene is something more flexible: it is like a writer's paragraph, constantly being revised. Scenes are not captured (as in a photo) but built up incrementally. Layers of visual and audio refinement are added over a crude outline of the motion, the mix constantly in flux, always changeable. George Lucas's last *Star Wars* movie was layered up in this writerly way. He took the action "Jedis clashing swords—no background" and laid it over a synthetic scene of a bustling marketplace, itself blended from many tiny visual parts. Light sabers and other effects were digitally painted in later, layer by layer. In this way, convincing rain, fire and clouds can be added in additional layers with nearly the same kind of freedom with which Lucas might add "it was a dark and stormy night" while writing the script. Not a single frame of the final movie was left untouched by manipulation. In essence, a digital film is written pixel by pixel.

The recent live-action feature movie *Speed Racer,* while not a box-office hit, took this style of filmmaking even further. The spectacle of an alternative suburbia was created by borrowing from a database of existing visual items and assembling them into background, midground and foreground. Pink flowers came from one photo source, a bicycle from another archive, a generic house roof from yet another. Computers do the hard work of keeping these pieces, no matter how tiny and partial they are, in correct perspective and alignment, even as they move. The result is a film assembled from a million individual existing images. In most films, these pieces are handmade, but increasingly, as in *Speed Racer,* they can be found elsewhere.

In the great hive-mind of image creation, something similar is already happening with still photographs. Every minute, thousands of photographers are uploading their latest photos on the Web site Flickr. The more than three billion photos posted to the site so far cover any subject you can imagine; I have not yet been able to stump the site with a request. Flickr offers more than 200,000 images of the Golden Gate Bridge alone. Every conceivable angle, lighting condition and point of view of the Golden Gate Bridge has been photographed and posted. If you want to use an image of the bridge in your video or movie, there is really no reason to take a new picture of this bridge. It's been done. All you need is a really easy way to find it.

Similar advances have taken place with 3D models. On Google SketchUp's 3D Warehouse, you can find insanely detailed three-dimensional virtual models of most major building structures of the world. Need a street in San Francisco? Here's a filmable virtual set. With powerful search and specification tools, high-resolution clips of any bridge in the world can be circulated into the common visual dictionary for reuse. Out of these ready-made "words," a film can be assembled, mashed up from readily available parts. The rich databases of component images form a new grammar for moving images.

After all, this is how authors work. We dip into a finite set of established words, called a dictionary, and reassemble these found words into articles, novels and poems that no one has ever seen before. The joy is recombining them. Indeed it is a rare author who is forced to invent new words. Even the greatest writers do their magic primarily by rearranging formerly used, commonly shared ones. What we do now with words, we'll soon do with images.

For directors who speak this new cinematographic language, even the most photo-realistic scenes are tweaked, remade and written over frame by frame. Filmmaking is thus liberated from the stranglehold of photography. Gone is the frustrating method of trying to capture reality with one or two takes of expensive film and then creating your fantasy from whatever you get. Here reality, or fantasy, is built up one pixel at a time as an author would build a novel one word at a time. Photography champions the world as it is, whereas this new screen mode, like writing and painting, is engineered to explore the world as it might be.

But merely producing movies with ease is not enough for screen fluency, just as producing books with ease on Gutenberg's press did not fully unleash text. Literacy also required a long list of innovations and techniques that permit ordinary readers and writers to manipulate text in ways that make it useful. For instance, quotation symbols make it simple to indicate where one has borrowed text from another writer. Once you have a large document, you need a table of contents to find your way through it. That requires page numbers. Somebody invented them (in the 13th century). Longer texts require an alphabetic index, devised by the Greeks and later developed for libraries of books. Footnotes, invented in about the 12th century, allow tangential information to be displayed outside the linear argument of the main text. And bibliographic citations (invented in the mid-1500s) enable scholars and skeptics to systematically consult sources. These days, of course, we have hyperlinks, which connect one piece of text to another, and tags, which categorize a selected word or phrase for later sorting.

All these inventions (and more) permit any literate person to cut and paste ideas, annotate them with her own thoughts, link them to related ideas, search through vast libraries of work, browse subjects quickly, resequence texts, refind material, quote experts and sample bits of beloved artists. These tools, more than just reading, are the foundations of literacy.

If text literacy meant being able to parse and manipulate texts, then the new screen fluency means being able to parse and manipulate moving images with the same ease. But so far, these "reader" tools of visuality have not made their way to the masses. For example, if I wanted to visually compare the recent spate of bank failures with similar events by referring you to the bank run in the classic movie *It's a Wonderful Life,* there is no easy way to point to that scene with precision. (Which of several sequences did I mean, and which part of them?) I can do what I just did and mention the movie title. But even online I cannot link from this sentence to those "passages" in an online movie. We don't have the equivalent of a hyperlink for film yet. With true screen fluency, I'd be able to cite specific frames of a film, or specific items in a frame. Perhaps I am a historian interested in oriental dress, and I want to refer to a fez worn by someone in the movie *Casablanca.* I should be able to refer to the fez itself (and not the head it is on) by linking to its image as it "moves" across many frames, just as I can easily link to a printed reference of the fez in text. Or even better, I'd like to annotate the fez in the film with other film clips of fezzes as references.

With full-blown visuality, I should be able to annotate any object, frame or scene in a motion picture with any other object, frame or motion-picture clip. I should be able to search the visual index of a film, or peruse a visual table of contents, or scan a visual abstract of its full length. But how do you do all these things? How can we browse a film the way we browse a book?

It took several hundred years for the consumer tools of text literacy to crystallize after the invention of printing, but the first visual-literacy tools are already emerging in research labs and on the margins of digital culture. Take, for example, the problem of browsing a feature-length movie. One way to scan a movie would be to super-fast-forward through the two hours in a few minutes. Another way would be to digest it into an abbreviated version in the way a theatrical-movie trailer might. Both these methods can compress the time from hours to minutes. But is there a way to reduce the contents of a movie into imagery that could be grasped quickly, as we might see in a table of contents for a book?

Academic research has produced a few interesting prototypes of video summaries but nothing that works for entire movies. Some popular Web sites with huge selections of movies (like porn sites) have devised a way for users to scan through the content of full movies quickly in a few seconds. When a user clicks the title frame of a movie, the window skips from one key frame to the next, making a rapid slide show, like a flip book of the movie. The abbreviated slide show visually summarizes a few-hour film in a few seconds. Expert software can be used to identify the key frames in a film in order to maximize the effectiveness of the summary.

The holy grail of visuality is to search the library of all movies the way Google can search the Web. Everyone is waiting for a tool that would allow them to type key terms, say "bicycle + dog," which would retrieve scenes in any film featuring a dog and a bicycle. In an instant you could locate the moment in *The Wizard of Oz* when the witchy Miss Gulch rides off with Toto. Google can instantly pinpoint desirable documents out of billions on the Web because computers can read text, but computers are only starting to learn how to read images.

It is a formidable task, but in the past decade computers have gotten much better at recognizing objects in a picture than most people realize. Researchers have started training computers to recognize a human face. Specialized software can rapidly inspect a photograph's pixels searching for the signature of a face: circular eyeballs within a larger oval, shadows that verify it is spherical. Once an algorithm has identified a face, the computer could do many things with this knowledge: search for the same face elsewhere, find similar-looking faces or substitute a happier version.

Of course, the world is more than faces; it is full of a million other things that we'd like to have in our screen vocabulary. Currently, the smartest object-recognition software can detect and categorize a few dozen common visual forms. It can search through Flickr photos and highlight the images that contain a dog, a cat, a bicycle, a bottle, an airplane, etc. It can distinguish between a chair and sofa, and it doesn't identify a bus as a car. But each additional new object to be recognized means the software has to be trained with hundreds of samples of that image. Still, at current rates of improvement, a rudimentary visual search for images is probably only a few years away.

What can be done for one image can also be done for moving images. Viewdle is an experimental Web site that can automatically identify select celebrity faces in video. Hollywood postproduction companies routinely "read" sequences of frames, then "rewrite" their content. Their custom software permits human operators to eradicate wires, backgrounds, unwanted people and even parts of objects as these bits move in time simply by identifying in the first frame the targets to be removed and then letting the machine smartly replicate the operation across many frames.

The collective intelligence of humans can also be used to make a film more accessible. Avid fans dissect popular movies scene by scene. With maniacal attention to detail, movie enthusiasts will extract bits of dialogue, catalog breaks in continuity, tag appearances of actors and track a thousand other traits. To date most fan responses appear in text form, on sites like the Internet Movie Database. But increasingly fans respond to video with video. The Web site Seesmic encourages "video conversations" by enabling users to reply to one video clip with their own video clip. The site organizes the sprawling threads of these visual chats so that they can be read like a paragraph of dialogue.

The sheer number of user-created videos demands screen fluency. The most popular viral videos on the Web can reach millions of downloads. Success garners parodies, mashups or rebuttals—all in video form as well. Some of these offspring videos will earn hundreds of thousands of downloads themselves. And the best parodies spawn more parodies. One site, TimeTube, offers a genealogical view of the most popular videos and their descendants. You can browse a time line of all the videos that refer to an original video on a scale that measures both time and popularity. TimeTube is the visual equivalent of a citation index; instead of tracking which scholarly papers cite other papers, it tracks which videos cite other videos. All of these small innovations enable a literacy of the screen.

As moving images become easier to create, easier to store, easier to annotate and easier to combine into complex narratives, they also become easier to be remanipulated by the audience. This gives images a liquidity similar to words. Fluid images made up of bits flow rapidly onto new screens and can be put to almost any use. Flexible images migrate into new media and seep into the old. Like alphabetic bits, they can be squeezed into links or stretched to fit search engines, indexes and databases. They invite the same satisfying participation in both creation and consumption that the world of text does.

We are people of the screen now. Last year, digital-display manufacturers cranked out four billion new screens, and they expect to produce billions more in the coming years. That's one new screen each year for every human on earth. With the advent of electronic ink, we will start putting watchable screens on any flat surface. The tools for screen fluency will be built directly into these ubiquitous screens.

With our fingers we will drag objects out of films and cast them in our own movies. A click of our phone camera will capture a landscape, then display its history, which we can use to annotate the image. Text, sound, motion will continue to merge into a single intermedia as they flow through the always-on network. With the assistance of screen fluency tools we might even be able to summon up realistic fantasies spontaneously. Standing before a screen, we could create the visual image of a turquoise rose, glistening with dew, poised in a trim ruby vase, as fast as we could write these words. If we were truly screen literate, maybe even faster. And that is just the opening scene.

Luke O'Brien

Spore's Intelligent Designer

Will Wright's new hit game is all about evolution. Or is it?

Game designer Will Wright has never been one for half measures. Wright's first hit, SimCity, released in 1989, set out to model the complexities of urban planning. Two decades later, he's moved on to a grander project. Wright's latest endeavor, Spore, tackles nothing less than life itself. You start with a single cell. Play long enough, and you'll evolve into an entire spacefaring society.

Transforming a blob of protozoa into a flock of Yuri Gagarins feels like a duty reserved for the almighty or, perhaps, epochal time. But it's nothing new for Wright. In the early 1990s, he released SimEarth and SimLife, precursors to Spore that covered similar ontological ground, putting the user in charge of developing life and planets. And then there was The Sims, the best-selling computer game franchise of all time, a virtual dollhouse that let you control the daily activities of cyberhumans. Wright has a knack for turning complex stuff into easily digestible entertainment. His new game, however, traffics in subjects most universities have multiple Ph.D. departments studying. By taking on evolution—and, by default, intelligent design—Spore wades into a roiling ecosystem.

So what happens when something as complex as human interaction or evolution gets reduced to mouse clicks? Naturally, distilling real life into a video game requires some simplifications. Wright, both by design and by necessity, takes artistic license with the intricate systems he models. His unique aesthetic sense has made him wildly successful. At the same time, it's turned several of his games into a battlefield for banner-waving geeks who are perpetually at loggerheads over the artist's agenda.

Take SimCity, in which players engage in municipal tasks such as zoning property, laying out power grids and streets, building police stations, and managing transportation. There's no city council or finicky court system. You play mayor, urban planner, and puppet master all at once—Rudy Giuliani's executive utopia.

While most SimCity addicts were busy building cities and then destroying them via earthquake, wonkier types were puzzling over the game's rules and value system. In a 1994 article in the *American Prospect,* Paul Starr referred to SimCity's "hidden curriculum." He noted that success required players to build cities on an industrial base, and he criticized the game's bias against mixed-use development. Private land values were pegged to the public budget, and the city's health depended on zoning and allocation of resources, which determined tax receipts. The underlying structure of the game was, in the words of Wright himself, a "capitalistic land value ecology."

Other critics questioned the absence of race, pointing out that simulating urban decay without taking ethnicity into account was unrealistic, if not manipulative. And then there were taxes. Raise them enough, and your citizens would riot. Every kid who played SimCity absorbed the underlying message: Taxes are dangerous. This was Milton Friedman in code. Still, it wasn't enough to satisfy conservatives. They said the game punished players for buying nuclear power plants while rewarding them for building mass transit. They grumbled that the game ignored the private market and depicted the state as the sole engine for urban growth. (For what it's worth, in the last year Wright has donated nearly $100,000 to Republican political causes. He backed Giuliani for president. He now supports McCain.)

The Sims, which came out in 2000, steered clear of policy issues. It did, however, raise questions about how to boil human behavior down to bytes. By the time The Sims 2 rolled out, the game's virtual inhabitants, who had different genetic backgrounds, could breed. This added heredity to the mix, and players learned to modify racial makeup and DNA. Some users even conducted studies in population genetics inside the game, tracking recessive and dominant alleles over generations.

The Sims, however, was never about modeling the descent of man or even human relationships. It focused mainly on the bureaucracy of life—the daily chores needed to keep your SimHuman from devolving into a slovenly, bankrupt outcast. But as usual with Wright's games, the approach didn't come without controversy. You don't make friends in The Sims—you acquire them. The more goods you amass, the more popular you become. The bigger the TV you stuff into your suburban palace, the happier you are.

"The constraints of consumer capitalism are built into the game's logic," wrote Ann McGuire, an Australian academic, echoing earlier complaints about the hypercapitalist SimCity. "The Sims distils and intensifies, through its underlying code, key ideological aspects of late capitalism: self, other, and time are all quantified and commodified. What the player is doing is shopping effectively in order to manage a life in the world."

It's hardly surprising, then, that Spore would be destined to provoke. Wright initially dubbed the game "SimEverything" because of the range of material it would cover. Months before the game came out, people started clucking on Internet forums. Would Spore take a scientific approach to evolution? Would it celebrate the tenets of intelligent design? Knowing Wright's history, it's no surprise that the answer is yes on both counts. It just depends whom you ask.

In Spore, players guide life through five different stages. Only the first two deal with evolution. You start as a cell, swimming around in a nutrient swamp, gobbling nourishment. The decisions you make from the start—whether to eat meat or plants or both, for example—set the course for your early development. As you progress, you earn "DNA points," opening up palettes of biological tweaks. Flagella help you swim faster. Spikes offer protection. That's evolution. But it's also where some people may see a divine hand. As the deity in this god game, your choices influence the game's outcome.

Some pro-I.D. groups have already targeted Spore as a possible educational vehicle. "It raises a lot of the questions we've been thinking about," Casey Luskin of the Intelligent Design Evolution and Awareness Center told me three months ago. "It has interesting pro-I.D. implications. . . . I know of at least two video-game developers affiliated with this who are pro-I.D." Luskin wouldn't tell me who those developers were, but he did recently weigh in on the Discovery Institute's blog to list five reasons why Spore will destroy common objections to intelligent design. His conclusion: "Spore is a video game that is intelligently designed to allow users to create fantasy worlds where evolution really can take place." (If a game that lets you play god is intelligently designed, does that make Will Wright some kind of deity? Could he be Audumbla, the icy cow of Norse legend that spawned the first gods by licking hoar frost?)

Spore's I.D. themes become more noticeable when you move onto land and into the "creature stage." Your goal here is to attain sentience. Your brain grows as you progress, interacting with other species through socialization, predation, or both. Particular behaviors put you on a path that opens up certain body parts. The range of options in the "creature creator" allows for an enormous variety of life—not as much as in nature, but a nice approximation. As it happens, intelligent design is good fun: You can spend hours with your critters, arranging spinal columns, attaching wings, and painting on polka dots. Or you can marvel at what other intelligent designers have dreamed up. Electronic Arts released the creature creator in June, and people have already cooked up millions of species, some elaborate and others obscene. (EA uploads your creations to servers and downloads other users' content into your world.)

At the same time, it's clear that Wright researched evolution. He appeared in a National Geographic Channel documentary called *How To Build a Better Being,* talking Darwin with evolutionary biologists and poring over fossils with paleontologists. He also consulted scientists who seem delighted, if mildly concerned, that their complex work is being simplified so dramatically. "Playing the game, you can't help but feel amazed how, from a few simple rules and instructions, you can get a complex functioning world with bodies, behaviors, and whole ecosystems," said Neil Shubin, a paleontologist at the University of Chicago. Just as Casey Luskin thinks Spore could get people excited about intelligent design, some biologists think the game could have educational value just by making users think about science, like an entertaining hook into evolutionary biology.

But the science in the game is wafer thin. Despite some overenthusiastic prognostications in reviews—"Spore could be the greatest gaming tool ever created to disseminate Darwinistic ideas," says one critic—the game makes no room for random mutation, the real source of differentiation. And natural selection plays only a minor role. If you don't bless your beast with a mouth or hands, you won't fare well. Almost anything else goes. At one point, my creature's legs and arms were connected by useless and mechanically impossible minilimbs. I did just fine. In Darwin's world, I would have been a snack for a more efficient predator.

Once the evolution stages end, Spore morphs into a traditional and less-innovative strategy game. You form a tribe, then evolve into a civilization with a military, economic, or religious culture. I managed to go religious by doing exactly what the religious nuts in America do not: eating lots of veggies and playing nice with my neighbors. When I eventually founded a city, I flooded the planet with religious propaganda to forcibly convert the unwashed heathens beyond my walls. This element of the game has angered atheists. I can't imagine that it's going to make evangelicals too happy, either.

So it goes with Wright. He admits Spore is a game that deals with intelligent design. He acknowledges the religious component. But he takes pains to point out that it's a caricature of reality, like all his games. The final stage of Spore has you scooting around in a spaceship, exploring a universe populated with user-created content. That's maybe not so realistic, but it is enjoyable. It's important to remember that building a game based strictly on evolutionary principles would be a disaster. How would you play it? Perhaps you'd just end up watching a lab computer churning data.

What people see as agendas in Spore and The Sims and SimCity may merely be artifacts of what's required to turn a simulation into a game. An early prototype of Spore included mutations, but Wright said it wasn't engaging—users needed to make those tweaks. "When we put the players in the role of intelligent designer then people were much more emotionally attached to what they made," he says.

Ultimately, games are made to engage the people who play them. Provoking wonderment or debate is a good thing.

Wright abstracts grandiose topics, and he does it well. Not enough game designers have the stones or the vision to try the same, which is why we get battered with endless versions of Madden NFL (also put out by Electronic Arts). In the end, that's also why Spore leaves such an impression. It's more than just fun. It's worth arguing about.

Adam Sternbergh

The Spreadsheet Psychic

Nate Silver is a number-crunching prodigy who went from correctly forecasting baseball games to correctly forecasting presidential primaries—and perhaps the election itself. Here's how he built a better crystal ball.

In a month when the Dow had its worst single-day plunge in over twenty years, when Lehman imploded, AIG faltered, and WaMu failed, when the word *crisis* became an everyday staple in newspaper headlines and the presidential race pulled close, then pulled apart, when the Chicago Cubs kicked off a playoff quest to win their first championship in 100 years (then got swept out in three straight games) and, for good measure, some scientists in an underground lab near the Swiss Alps fired up a Large Hadron Collider that some serious observers warned might create a black hole that would swallow up the Earth, it was comforting to sit down and have lunch in midtown with a man who can see the future. It's not that Nate Silver is psychic, or even that he's right all the time. He's just proved very good, especially of late, at looking at what's already happened and using that information to predict what will happen next.

Silver, who's 30, thin, and lives in Chicago, had been flown to New York at the invitation of a hedge fund to give a talk. "They just said, 'Why don't you come in, talk about your models,'" he said with a shrug. "I'll probably just take a lot of questions." Silver doesn't know all that much about high finance; these days, he's spending most of his energy on his political Website, FiveThirtyEight (the total number of Electoral College votes), where he uses data analysis to track and interpret political polls and project the outcome of November's election. The site earned some national recognition back in May, during the Democratic primaries, when almost every other commentator was celebrating Hillary Clinton's resurgent momentum. Reading the polls, most pundits predicted she'd win Indiana by five points and noted she'd narrowed the gap with Obama in North Carolina to just eight.

Silver, who was writing anonymously as "Poblano" and receiving about 800 visits a day, disagreed with this consensus. He'd broken the numbers down demographically and come up with a much less encouraging outcome for Clinton: a two-point squeaker in Indiana, and a seventeen-point drubbing in North Carolina. On the night of the primaries, Clinton took Indiana by one and lost North Carolina by fifteen. The national pundits were doubly shocked: one, because the results were so divergent from the polls, and two, because some guy named after a chili pepper had predicted the outcome better than anyone else.

Silver's site now gets about 600,000 visits daily. And as more and more people started wondering who he was, in May, Silver decided to unmask himself. To most people, the fact that Poblano turned out to be a guy named Nate Silver meant nothing. But to anyone who follows baseball seriously, this was like finding out that a guy anonymously running a high-fashion Website turned out to be Howard Cosell. At his day job, Silver works for Baseball Prospectus, a loosely organized think tank that, in the last ten years, has revolutionized the interpretation of baseball stats. Furthermore, Silver himself invented a system called PECOTA, an algorithm for predicting future performance by baseball players and teams. (It stands for "player empirical comparison and optimization test algorithm," but is named, with a wink, after the mediocre Kansas City Royals infielder Bill Pecota.) Baseball Prospectus has a reputation in sports-media circles for being unfailingly rigorous, occasionally arrogant, and almost always correct.

This season, for example, the PECOTA system predicted that the Tampa Bay Rays would win 90 games. This seemed bold, even amusing, given that the Rays were arguably the worst team in baseball. In 2007, they'd *lost* 96 games. They'd finished last in all but one season of their ten-year existence. (In 2004, they finished fourth.) They had some young talent, sure, but most people, even those in the Rays' front office, thought that if the team simply managed to win more games than it lost, that would represent a quantum leap.

PECOTA, however, saw it differently. PECOTA recognized that the past Rays weren't a hopelessly bad team so much as a good team hampered by a few fixable problems—which, thanks to some key off-season changes, had been largely remedied. Silver argued on the Baseball Prospectus Website that the long-suffering team had finally "decided to transform themselves from a sort of hedge fund for undervalued assets into a real, functional baseball club."

PECOTA, as it turns out, wasn't exactly right. The Rays didn't win 90 games this year. They won 97 games and are currently playing the Red Sox for the American League championship.

So, Nate Silver: What happens next?

Sports and politics offer several obvious parallels. Both involve a competition, essentially between two teams. Both involve reams of statistical data available for devotees to sort through

—or, more commonly, for intermediary experts to sort through, analyze, and then interpret for you. In baseball, these stats track player performance—how many hits a player gets, and when, and against what kind of pitchers—while in politics, the data tracks voter preferences. Who do you like and why? What kind of choice are you likely to make on Election Day? These stats, on their face, seem pretty straightforward. If a hitter hits .300, he's valuable. If Obama opens up a six-point national lead, he's in good shape.

And yet in both sports and politics, there's an industry built around studying this data, making up stories about it, and then trying to sell those stories to you. For example: A-Rod, for all his greatness, can't deliver in the clutch. Obama, for all his charisma, has struggled to connect with white working-class voters. The Mets are a bunch of chokers. This election, it's all about the hockey moms.

As a result, in baseball and, now, politics, there exists a small subculture of counterexperts: People who argue against these conventional story lines using new interpretations of the raw data to make their case. In baseball, this counterculture has been growing for roughly 30 years and can be traced, improbably, to one man: Bill James, a cranky Midwesterner who started writing articles about baseball while working the night shift as a security guard at a pork-and-beans factory in Lawrence, Kansas. In 1977, he published a photocopied newsletter called *Baseball Abstract,* which found a cult following that later blossomed into a national audience. By the late eighties, he was hailed as the founder of "sabermetrics"—a new field dedicated to better analysis of baseball stats—and the father of a revolution. Once considered a Unabomber-style outcast, James now consults for the Red Sox.

And in the nineties, the study of sabermetrics exploded. For starters, the development of so-called fantasy baseball—a game in which fans draft a team of real players, then compete with each other based on the players' on-field success—created a huge new market for performance projections. (If you want to win your fantasy league, you care a lot less about who hit 40 home runs last year than you do about who'll hit 40 next year.) And the advent of the Internet allowed fans unprecedented access to stats, both raw and packaged by various experts. Then, in 2003, Michael Lewis wrote *Moneyball,* a best-selling book that valorized Billy Beane, the general manager of the Oakland A's, for using some of these new insights to overcome the financial advantage of richer teams.

In the midst of all this, in 1996, Baseball Prospectus was born. Founded by five baseball fans who met each other online, the BP crew are like the bratty children of Bill James, adding a new level of analytical sophistication to his contrarian philosophy. "When he started, Bill James had to actually call up teams and ask for their information," says Joe Sheehan, one of BP's founders. "Now we're able to download databases. We can do things with one-tenth the effort and a hundred times the available data." Also, whereas James used stats to explain what a player had done, BP uses stats to predict what a player might do. As a result, BP has built a small but successful empire of smarty-pants, with a Website, syndicated columns, and most prominently, a preseason annual full of player projections.

As an avid fantasy player, I've spent long hours combing through the pages of the Baseball Prospectus book. It arrives each season huge, heavy, and intimidatingly dense. ("It's longer than *Moby-Dick* and heavier than *War and Peace,*" jokes Steven Goldman, one of the editors.) Last year's edition weighed in at 605 oversize pages and offered essays like "Expanding the Cannon: Quantifying the Impact of Outfield Throwing Arms." The writing is lively and funny, nerd-nip for baseball obsessives. But the book, with its extensive charts and graphs, its talk of VORP (value over replacement player) and

SNLVAR (support-neutral lineup adjusted value added above replacement), its SAC percent and EqH9 (oh, never mind), can also make you feel like you're reading a Ph.D. dissertation in statistics or a book by Dr. Seuss—or both, at the same time.

In fact, the work of stat hounds in general, and of Baseball Prospectus in particular, is so obviously the product of high-wattage brainpower and creativity that you can't help but occasionally wonder: *What if someone applied all this energy to something that actually mattered, like, I don't know, politics?*

Last year, at the start of an unusually unpredictable election season, Nate Silver began to wonder the same thing.

As stats are to baseball, polls are to politics; i.e., the basic numeric measurement of how things have gone in the past and how they might go in the future. Ask any pollster, though, and he will tell you that polls aren't meant to be used as predictive tools—they're simply a rough measure of where the electorate stands at a given moment. As pollster John Zogby put it to me, "We take snapshots. And when you take many snapshots in a row, you get motion pictures."

But unlike baseball stats, polls are a notoriously imprecise measurement. In baseball, at least, a hit is a hit. With polls, a yes isn't always a yes. Sometimes it's more like a "maybe," or a "yes, until I change my mind," or an "I don't know, but I'll say yes anyway to get you off the phone." Poll results can vary dramatically based on what you're asking, who you're asking, how you're asking, and how many people decide to answer you. Three different polls were conducted recently asking Americans how they felt about the federal $700 billion bailout. They all asked the question in slightly different ways and the results were essentially useless: One poll had people in favor of the bailout 57 to 30 percent, one had them against it 55 to 31, and one was basically split down the middle. In other words, polls are, at best, educated guesses. But if there's one thing Nate Silver loves to make, it's an educated guess.

In this year's Democratic primary, for example, the polls were all over the place. Before the Iowa caucuses on January 3, one poll had Clinton winning by nine, one had Clinton by two, and one had Obama by one. Obama won by seven. In the New Hampshire primary, five days later, one poll had Obama by thirteen and most others had him winning by eight or nine. Clinton won by three. Primaries are notoriously difficult to poll, because unlike in a general election, turnout is very unpredictable and people are much more likely to switch their choice at the last minute. As the primaries went on, however, Silver, who had been writing an anonymous diary for the liberal Website Daily Kos, made an observation about this year's voters: While the polls were wobbling wildly state-to-state, the demographic groups supporting each candidate, and especially Clinton and Obama, were remarkably static. He wasn't the only one who noticed this, of course—it was a major narrative theme of the campaign. One pundit summed it up by saying that Clinton had "the beer track"—blue-collar whites, Latinos, and seniors—while Barack had African-Americans and "the wine track": young voters and educated whites.

Every other pundit, though, was doing what they've always done, i.e., following the polls. Silver decided to ignore the polls. Instead, he used this observation about demographics to create a model that took voting patterns from previous primaries and applied them to upcoming contests. No phone calls, no sample sizes, no guesswork. His crucial assumption, of course, was that each demographic group would vote in the same way, in the same percentages, as they had in other states in the past.

Like many of the so-called *Moneyball* breakthroughs in baseball, this was both a fairly intuitive conclusion and a radical break from conventional thinking. (In *Moneyball,* for example, the idea that players who get on base most often are the most valuable—which now seems kind of obvious—was a major breakthrough in strategy.) After all, political pundits love to talk about states as voting blocs—New Hampshire's leaning this way, North Carolinians care about this, etc.—as though residency is the single most important factor in someone's vote. Silver's model more or less ignored residency. But his hunch about demographics proved correct: It's how he called the Indiana and North Carolina results so accurately when the polls got them so wrong.

In baseball, a hit is a hit. With polls, a yes isn't always a yes. Sometimes it's more like "I don't know, but I'll say yes anyway to get you off the phone."

The model didn't always work throughout the primaries: Silver missed on Kentucky and South Dakota. But the model proved that the kind of creative swashbuckling that exemplifies Baseball Prospectus—the institutional obsession with questioning assumptions, even your own, even (or especially) to the point of heresy—could work when applied to politics as well. When I asked Joe Sheehan to sum up the Baseball Prospectus philosophy, he said simply, "Back up your argument. Because too many people are telling stories, as opposed to actually looking for the truth."

Meanwhile, even as his primary model attracted attention, Silver was cooking up another idea. He figured there must be a better way to use the daily tracking polls to predict a candidate's future, just as he'd once found a better way to use baseball stats to predict how many home runs a player might hit. His simple goal, as he explained on Daily Kos in late February, was to "assess state-by-state general-election polls in a probabilistic manner." In other words, he wanted to find a way to use all those occasionally erratic, occasionally unreliable, occasionally misleading polls to tell him who would win the election in November, which at that point was over 250 days away.

It's a tough business, being an oracle. Everyone cheers when you hit a bull's-eye, but no one's arrows fly true all the time. "Sometimes being more accurate means you're getting things right 52 percent of the time instead of 50," says Silver. "PECOTA is the most accurate projection system in baseball, but it's the most accurate by half a percent." That half-percent, though, makes all the difference. Silver's work, in both baseball and politics, is about finding that slim advantage. "I hate the first 90 percent [of a solution]," he says. "What I want is that last 10 percent."

As a kid, Silver was not a dork in a plastic bubble, as you might expect, gobbling stats and spouting figures. He grew up in East Lansing, Michigan, a typical baseball fan with a Tigers pennant on his bedroom wall. In person, talking baseball, he hardly comes off as a human computer; rather, he talks with the same bursts of enthusiasm familiar to any engaged fan in a sports bar. (And it's been a rough year for Silver, fan-wise: His home team, the Tigers, were an underperforming disaster, and his two adopted teams, the Cubs and the White Sox, were both quickly and tragically dispatched from the playoffs this year.)

His approach to politics is similar—he's an engaged fan. He unapologetically roots for Obama. One of his early posts as a contributor to Daily Kos, titled "I Got Dinner With Barack Obama," recounts with gee-whiz wonder a chance sighting of Obama during Silver's birthday outing at a Mexican restaurant. ("At first I was pissed off with my friend for not doing more to alert me," he wrote, "but if I'd had more advance warning, I'd probably have done something stupid like scream 'Fired up!,' which would have been embarrassing in retrospect.") But he doesn't try to pummel you with numbers to prove his argument, like a typical hot-blooded partisan. Instead, on his site, he exhibits the cool confidence of someone who's simply used to knowing his stuff better than anyone else in the room.

Not that he can't pick, and win, a good fight. In a post on Daily Kos last December titled "Is a Bad Poll Better Than No Poll at All?" Silver singled out a few pollsters, particularly American Research Group, to show that their consistently off-base numbers will skew polling averages so severely that they harm one's results. Of ARG he wrote, "They have a track record of rolling out some polls that are completely different from anybody else in the race, and when they do, they are almost always wrong."

Dick Bennett, the pollster for ARG, responded by posting items on his Website such as "Nate Silver is Wrong Again," and mocking FiveThirtyEight's slogan ("Electoral Projections Done Right") in a tone that echoed current political attack ads. ("So much for electoral projections done right.") Then, in June, Silver posted an open letter to Bennett, which read, "It has been a long and hard-fought primary campaign. We've both had our share of successes, and made our share of mistakes. Granted, you made a few more than I did" —and in that last sentence, every word but "Granted" was a separate link to an ARG polling misfire. Then Silver challenged Bennett to a contest, in which each site would call the elections results, state-by-state, with a $1,000 bounty per state. Bennett never took him up on it, and this is what he has to say about Silver now: "What he does is different than what I do. There's a market for that. There's also a lady down the street who will read your palm."

As a high-schooler, Silver was a state-champion debater, though he claims to be only a so-so public speaker. I asked if he ever thought of becoming a baseball G.M., like the 34-year-old boy wonder (and sabermetrics proponent) Theo Epstein, who took the Red Sox to two championships (and counting). Silver said, "The people who do that are very talented. They're very smart, very polished. And I'm not much of a schmoozer. With Baseball Prospectus, you still have a voice and it's influential. I prefer shaping public opinion, I suppose."

After earning a degree in economics from the University of Chicago, Silver took a corporate job at a consulting firm but found it boring. He seems endlessly distractable; for example, in 2007, he started a Website, The Burrito Bracket, that rated Mexican restaurants in Wicker Park. ("Each week, I will be visiting two restaurants and having the same item of food [carne asada burritos, for example] at each one. The restaurant that provides the superior experience advances to the next round of the bracket.") For a while, he was supplementing his income playing online poker, and even earned six figures one year, but eventually he quit. "For a while, there was a lot of money to be made, but you kind of eliminate one sucker at a time," he says, "until finally you're the sucker." After he developed PECOTA and joined Baseball Prospectus, he turned his eye to political analyses, thus finding another field in which to identify suckers and eliminate them one at a time.

In concocting FiveThirtyEight, Silver decided the best way to read the polls was to put them all together, with the idea that averaging ten polls would give you a better result than trying to pick out the best one. Again, he wasn't the first person to do this—other sites like RealClearPolitics and Pollster offer the same service. But, as Silver told me, "Sometimes the answer is in looking at other alternatives that exist in the market and saying, 'They have the right idea, but they're not doing it quite the right way.'"

Silver wanted to average the polls, but he wanted the polls that were more accurate to count for more, while the bad polls would be discounted. Other sites, like RealClearPolitics, treat every poll as equal. Silver knows that some polls are simply better than others. Yet it's hard to know how accurate a general-election poll is before the actual election.

So he came up with a system that predicts a pollster's future performance based on how good it's been in the past. In finding his average, Silver weights each poll differently—ranking them according to his own statistic, PIE (pollster-introduced error)—based on a number of factors, including its track record and its methodology. One advantage of this system is that, during the primaries, the system actually got smarter. Because each time a poll performed well in a primary, its ranking improved.

For the general election, this gets trickier, since you have polls coming out every single day and you can't know which ones are getting it right until Election Day. You can, however, weigh these new polls based on the pollster's history, the poll's sample size, and how recently the poll was conducted. You can also track trends over time and use these trend lines to forecast where things will end up on November 4. You can also, as Silver has done, analyze all the presidential polling data back to 1952, looking for information as to what is likely to happen next. (For example, how much the polls are likely to tighten in the last month of the race, which they traditionally do.) You can also run 10,000 computer simulations of the election every day based on your poll projections. (Think of this as sort of like that scene at the end of *WarGames,* where the computer blurs through every possible nuclear-war scenario.) As of October 8, the day after the town-hall debate, Silver's simulations had Obama winning the election 90 percent of the time.

All of which is very seductive (and heartening to Obamaphiles), especially when you see it laid out on Silver's site, with its pleasing graphs, compelling charts, and graphically vibrant electoral maps. But in essence, Silver's whole undertaking is premised on breaking the first rule of reading polls: He's assuming—in his complex, elegant, partially proprietary and yet-to-be-entirely-validated way—that today's polls can predict tomorrow's election. Rival statisticians, in particular Sam Wang at Princeton (who runs his own poll-aggregation blog), have criticized Silver, arguing that polls can't and shouldn't be used as a crystal ball. Other critics have argued that the idea of a projection model is inherently flawed because it can't predict the unpredictable—for example, before the financial meltdown and McCain's campaign-suspension stunt, the polls were much tighter and Silver's electoral map had McCain on top.

Silver agrees, to a point, comparing daily polls, especially ones that come out months before the election, to "a 162-game baseball season, where one individual win or loss doesn't really tell you that much about the ultimate outcome." But for him, you have to use polls to predict the future—that's the whole point. Unlike other electoral projection maps, Silver calls each state for one candidate or the other—there are no undecideds—because the goal is to approximate what the map will look like after Election Day. "I think the entire value of the exercise is in predicting the outcome in November," Silver says, in a response to Wang. "What would happen in an election held today is a largely meaningless question."

And Silver is right. The truth is that everyone reads polls this way. When you pick up the paper and see McCain up by three or Obama up by six, you assume that means that candidate is on his way to a win. Silver's goal with FiveThirtyEight, then, is to simply do what everyone does, but do it better—to read the polls in such a way that those assumptions we all naturally make will actually turn out to be true.

For all the numbers and nuance, the adjustments and algorithms, there's really only one stark, looming, unambiguous test for the political prognosticator. "Pretty soon, there's going to be an Election Day," says pollster J. Ann Selzer of Selzer & Co. "And you're either going to be golden or a goat." (Selzer, this season, has consistently been golden, calling the Iowa caucus flawlessly, which is partly why Selzer & Co. is Silver's top-ranked pollster on FiveThirtyEight.)

Like everyone else calling this election, Silver's day of reckoning will come on November 4. In the meantime, he's become an increasingly confident commentator, growing into a national role as a calming anti-pundit among the white noise of partisan spin. FiveThirtyEight not only tracks numbers, but features field reports from the 50 states by Silver's colleague, Sean Quinn, and the documentary photographer Brett Marty. And Silver posts to his blog several times daily, spotting and dissecting surprising trends or aberrations, such as a recent poll in Minnesota that handed McCain a sudden one-point lead. Normally, you'd expect a liberal-leaning commentator to read such a result and blame bias, or error, or voodoo. Silver, however, poked around and determined that McCain had been recently outspending Obama three-to-one in Minnesota, making it the only state in the country where he was out-advertising Obama. "So McCain may literally have bought his way into a competitive race," Silver wrote. "So, yes, you can beat a state into submission if you really want to. . . . But whether it's been a *good* use of resources, we'll have to see."

I caught up with Silver on the phone recently, on a day when he'd just arrived back in Chicago from New York, having appeared the night before on *The Colbert Report*. We talked about Obama's widening lead over McCain, and the remaining undecideds—a group, he says, that's "mostly older rather than younger, more religious than not, and a lot of Independents, which is typical"—and how a lot of formerly solid red states now look like good bets to turn blue. "My pet theory is that these states along the Atlantic coast, like Virginia and North Carolina, are growing so fast that you have a lot of newly registered Democrats. That universe now contains people that it didn't a month ago. Literally, the composition of the North Carolina electorate is different than it was six months ago. They update their registration figures on a weekly basis, and last week the Democrats registered about 16,000 new voters—which would represent one percent of their turnout from 2004." And this year, at least, for all the surprises and *Sturm und Drang,* the electorate appears to be acting rationally. "The conventional punditry underestimates voters. The voters are pretty smart. They picked two very strong candidates. Even if McCain's in trouble now, if it had been Fred Thompson, he might have conceded already."

Even as he updates his projections and runs his 10,000 simulations a day, Silver wonders if maybe we don't yet know what the narrative of this election will be. In September, he wrote a post on Obama's extensive "ground game"—his efforts to set up outposts and register new voters, which have far outstripped McCain's—and suggested, "Suppose that, because of their ground efforts, the Obama campaign is 5 percent more efficient at turning out its vote than the McCain campaign on Election Day. . . . The implications of this would be enormous—a net of two to three points in each and every swing state—but we know zip, zilch, nada at this stage about their ultimate effect."

This is the paradoxical spirit of the stat-heads: They can be arrogant, sure, and even bullying as they charge forward, brandishing their spreadsheets. But they are just as happy to prove themselves wrong as they are to debunk anyone else. This, I think, is at the heart of Silver's appeal. (From a recent random Facebook status update: "I am an empiricist and I trust Nate Silver. Read it and chill.") "Nate's medium-term goal is to accomplish what we've accomplished at Prospectus —to change the conversation," says Sheehan. "And Nate's growth curve has been much sharper than ours ever was. He's crammed about five years' of BP's growth into five months. And if you get good enough arguments out there, if you do your work well enough, then other people have to do their work better. Nate's watched that at Prospectus. But FiveThirtyEight can do things for America that Baseball Prospectus never could."

FiveThirtyEight is the product of a movement, but also of a moment. The political media is polarized. Cable anchors choke on their own spin. The red states and blue states act like the Jets versus the Sharks—they don't trust us and we don't trust them. So we all rail against the enemy in the echo chambers of comment boards, retreating to the bomb-shelter safety of partisan blogs.

It's not that Silver is objective or impartial—he's not. He's still that young guy who almost yelled "Fired up!" across a crowded Mexican restaurant. But his ultimate goal is simple and nonpartisan: to build a better expert. Sure, he'll be disappointed if Obama loses. But he also says, "If Obama does lose, I think it's healthy to try and understand why, rather than just kicking and throwing things." If he ever decides to run for office, that wouldn't make for a bad slogan. Nate Silver: More understanding. Less kicking and throwing things.

Clay Shirky

Gin, Television, and Cognitive Surplus

Where's the mouse?

I was recently reminded of some reading I did in college, way back in the last century, by a British historian who was arguing that the critical technology for the early phase of the industrial revolution was gin. The transformation from rural to urban life was so sudden, and so wrenching, that the only thing society could do to manage was to drink itself into a stupor for a generation. The stories from that era are amazing—there were gin pushcarts working their way through the streets of London.

And it wasn't until society woke up from that collective bender that we actually started to get the institutional structures that we associate with the industrial revolution today. Things like public libraries and museums, increasingly broad education for children, elected leaders—a lot of things we *like*—didn't happen until having all of those people together stopped seeming like a crisis and started seeming like an asset.

It wasn't until people started thinking of this social density as a vast civic surplus, one they could design for rather than just dissipate, that we started to get what we think of now as an industrial society.

If I had to pick the critical technology for the 20th cen-

/herecomeseverybody.org/

tury, the bit of social lubricant without which the wheels would've come off the whole enterprise, I'd say it was the sitcom. Starting with the Second World War a whole series of things happened—rising GDP per capita, rising educational attainment, rising life expectancy and, crucially, a rising number of people who were working five-day weeks. For the first time, society forced an enormous number of its citizens to manage something they had never had to manage before—*free time.*

And what did we do with that free time? Well, mostly we spent it watching TV.

We did that for decades. We watched *I Love Lucy.* We watched *Gilligan's Island.* We watch *Malcolm in the Middle.* We watch *Desperate Housewives. Desperate Housewives* has essentially functioned as a kind of cognitive heat sink, dissipating thinking that might otherwise have built up and caused society to overheat.

It's only now, as we're waking up from *that* collective bender, that we're starting to see the cognitive surplus as an asset rather than as a crisis. We're seeing things being designed to take advantage of that surplus, and to deploy it in ways more engaging than just having a TV in everybody's basement.

This hit me in a conversation I had about two months ago. I had recently finished a book called *Here Comes Everybody,* which was just published, and this particular recognition came out of a conversation I had about the book. I was being interviewed by a TV producer to see whether I should be on their show, and she asked me, "What are you seeing out there that's interesting?"

I started telling her about the Wikipedia article on Pluto. You may remember that Pluto got kicked out of the planet club a couple of years ago, so all of a sudden there was all of this activity on Wikipedia. The talk pages light up, people are editing the article like mad, and the whole community is in a ruckus: "How should we characterize this change in Pluto's status?" And a little bit at a time they move the article—fighting offstage all the while—from describing Pluto as the ninth planet to describing it as an odd-shaped rock with an odd-shaped orbit at the edge of the solar system.

So I tell her all this stuff, and I think, "Okay, we're going to have a conversation about authority or social construction or whatever." But that wasn't where she was going. She heard this story and she shook her head and said, "*Where do people find the time?*" That was her question, and I just kind of snapped. I said, "No one who works in TV gets to ask that question. You know where the time comes from. It comes from the cognitive surplus you've been masking for 50 years."

So how big is that surplus? If you take Wikipedia as a kind of unit, all of Wikipedia, the whole project—every page, every edit, every talk page, every line of code, in every language that Wikipedia exists in—that represents something like 100 million hours of human thought. I worked this out with Martin Wattenberg at IBM; it's a back-of-the-envelope calculation, but it's the right order of magnitude: about 100 million hours of thought.

And television watching? Two hundred billion hours, in the U.S. alone, *every year*. Put another way, now that we have a unit, that's 2,000 potential Wikipedia projects a year spent watching television. Or put still another way, in the U.S., we spend 100 million hours every weekend just watching the ads. That's a pretty big surplus. People who look at projects like Wikipedia and ask, "Where do they find the time?" don't understand how tiny that entire project is, as a portion of this asset, this cognitive surplus that's finally being dragged into what Tim O'Reilly calls an architecture of participation.

Now, the interesting thing about a surplus like that is that society doesn't know what to do with it at first—hence the gin, hence the sitcoms. Because, of course, if people knew how to adapt that surplus to the existing social institutions, then it wouldn't be a surplus, would it? It's precisely when no one has any idea how to use the surplus effectively that people have to start experimenting, in order for the surplus to become integrated. And the course of that integration can transform society.

The early phase for taking advantage of this cognitive surplus, the phase I think we're still in, is full of special cases. The physics of participation is much more like the physics of weather than it is like the physics of gravity. We know all the forces that combine to make these kinds of things work: there's an interesting community over here, there's an interesting sharing model over there, those people are collaborating on open source software. But despite knowing the inputs, we can't predict the outputs yet because there's so much complexity.

The way to explore complex ecosystems is to just try lots and lots and lots of things, and hope that everybody who fails, fails informatively so that you can at least find a skull on a pikestaff near where you're going. That's the phase we're in now.

Here's one example—it's a tiny example but also one I especially love. A couple of weeks ago one of my students at NYU's Interactive Telecommunications Program forwarded me a project started by a professor in Brazil, in Fortaleza, named Vasco Furtado. It's a Wiki map for crime in Brazil. If there's an assault, if there's a burglary, if there's a mugging, a robbery, a rape, a murder, you can go and put a pushpin on a Google map, and you can characterize the assault, and you start to see a map of where these crimes are occurring.

Now, these coordinates already exist as tacit information. Anybody who knows a town has some sense of, *"Don't go there. That street corner is dangerous." "Don't go in this neighborhood." "Be careful there after dark."* But it's information society knows without society *really* knowing it, which is to say that there's no public source where you can take advantage of it. And the cops, if they have that information, are certainly not sharing it. In fact, one of the things Furtado says in starting the Wiki crime map was, "This information may or may not exist some place in society, but it's actually easier for me to try to rebuild it from scratch than to try and get it from the authorities who might have it now."

Maybe Furtado's project will succeed or maybe it will fail. The normal case of social software is still failure; most of these kinds of experiments simply don't pan out. But the ones that do are incredible—and, obviously, I hope that this one succeeds. Even if it doesn't, though, it's illustrated the main point already, which is that someone working alone, with really cheap tools, has a reasonable hope of carving out enough of the cognitive surplus, enough of the desire to participate, enough of the collective goodwill of the citizens, to create a resource you couldn't have imagined existing even five years ago.

So that's the answer to the question, "Where do they find the time?" Or, rather, that's the numerical answer. But that question prompted another observation as well. At some point in this same conversation with the TV producer, I talked about World of Warcraft guilds, and as I was talking, I could sort of see that she was thinking: "Losers. Grown men sitting in their basement pretending to be elves."

And I thought: *At least they're doing something.*

Did you ever see that episode of *Gilligan's Island* where they almost get off the island and then Gilligan messes up and so they don't? I saw that one. I saw that one *a lot* when I was growing up. And every half-hour that I watched it was a half an hour I wasn't posting at my blog or editing Wikipedia or contributing to a mailing list. Of course I had an ironclad excuse for not doing those things, which is that none of those things existed then. I was forced into the channel of media because it was the only one. But now it's not, and that's the big surprise. However lousy it is to sit in your basement and pretend to be an elf, I can tell you from personal experience, it's way worse to sit in your basement trying to figure who's cuter, Ginger or Mary Ann.

I'm willing to raise that idea to a general principle: *It's better to do something than to do nothing.* Even lolcats, cute pictures of kittens made even cuter with the addition of cute captions, hold out an invitation to participation. One of the things the lolcat says to the viewer is, "If you have some sans-serif fonts on your computer, you can play this game, too." And that message—*you can do this, too*—represents a big change.

This is something that people in the media world don't understand. Media in the 20th century was run as a single race—consumption. How much can we produce? How much can you consume? Can we produce more so that you'll consume more? And the answer to those questions has generally been yes. But media is actually a triathlon, composed of three different events. People like to consume, but they also like to produce, and they also like to share.

What's astonishing to people who were committed to the structure of the previous society, before we tried to take this surplus and do something interesting, is the discovery that people who are offered the opportunity to produce and to share will often take the offer. That doesn't mean that we'll never sit around mindlessly watching *Scrubs* on the couch. It just means we'll do it less.

And that's the other thing about the size of the cognitive surplus we're talking about. It's so large that even a small change could have huge ramifications. Let's say that everything stays 99 percent the same, that people watch 99 percent as much television as they used to, but 1 percent of that time is now carved out for producing and for sharing. The Internet-connected population watches roughly a *trillion* hours of TV a year. That's about five times the size of the annual U.S. consumption. One percent of that is 100 Wikipedia projects per year worth of participation.

I think that's going to be a big deal. Don't you?

Well, the TV producer did not think so; she was not digging that line of thought at all. And her final question to me was essentially, "Isn't this all just a fad?" You know, sort of the 21st-century equivalent of flagpole-sitting? Sure, it's fun to go out and produce and share a little bit, but eventually people are going to realize, "This isn't as good as doing what I was doing before," and settle down. And I made a spirited argument that no, this wasn't the case, that this was in fact a big one-time shift, closer to the industrial revolution than to flagpole-sitting.

I argued that this isn't the sort of thing society grows out of: It's the sort of thing that society *grows into.* But I'm not sure she believed me, in part because she didn't want to believe me, and in part because I didn't have the right story yet. And now I do.

I was having dinner with a group of friends about a month ago, and one of them was talking about sitting with his four-year-old daughter watching a DVD. In the middle of the movie, apropos nothing, his daughter jumps up off the couch and runs around behind the screen. It seemed like a cute moment. Maybe she went back there to see if the characters were really back there—or whatever. But that wasn't what she was doing. She started rooting around in the cables. And when her dad asked what she was doing, she stuck her head out from behind the screen and said, "Looking for the mouse."

Here's something four-year-olds know: A screen that ships without a mouse ships broken. Here's something four-year-olds know: Media that's targeted at you but doesn't include you may not be worth sitting still for. Those are things that make me believe that this is a one-way change. Because four-year-olds, the people who are soaking most deeply in the current environment, who won't have to go through the trauma that I have to go through of trying to unlearn a childhood spent watching *Gilligan's Island,* just assume that media includes consuming, producing and sharing.

So now I have a new answer when people ask me what we're doing (and when I say "we" I mean both the larger society trying to figure out how to deploy this cognitive surplus and also, and especially the people who are working hammer and tongs at figuring out the next good idea). From now on, I'm going to tell them: *We're looking for the mouse.* We're going to look at every place that a reader or a listener or a viewer or a user has been locked out of or has been served up a passive or a fixed or a canned experience, and we're going to ask ourselves, "If we carve out a little bit of the cognitive surplus and deploy it here, could we make a good thing happen?" And I'm betting the answer is yes.

About the Contributors

danah boyd, Ph.D., is a social media researcher at Microsoft Research and a fellow at Harvard University's Berkman Center for Internet and Society. Her research focuses on how American youth navigate new genres of social media in the course of their everyday lives. She also regularly blogs at www.zephoria.org/thoughts/.

Nicholas Carr is the author of *The Big Switch: Rewiring the World, from Edison to Google* and *Does IT Matter?* He has written for *The Atlantic,* the *New York Times, The Guardian, Wired,* and the *Financial Times,* among other periodicals.

Dalton Conley is University Professor and Dean of Social Sciences at New York University. He also holds appointments at NYU's Wagner School of Public Service, as an Adjunct Professor of Community Medicine at Mount Sinai School of Medicine, and as a Research Associate at the National Bureau of Economic Research (NBER). In 2005, Conley became the first sociologist to win the National Science Foundation's Alan T. Waterman Award. His research focuses on the determinants of economic opportunity within and across generations. He is the author of several books, including *Elsewhere, USA; Honky; The Pecking Order;* and *Being Black, Living in the Red: Race, Wealth, and Social Policy in America.* Conley lives in New York City.

Joshua Davis is a contributing editor at *Wired* magazine. His work previously appeared in the 2006 and 2007 editions of *The Best of Technology Writing.* His book *The Underdog* chronicles his efforts, among other things, to become the lightest man to sumo wrestle at the U.S. Sumo Open.

Julian Dibbell has, in the course of over a decade of writing and publishing, established himself as one of digital culture's most thoughtful and accessible observers. He is the author of two books about online worlds—*Play Money: Or How I Quit My Day Job and Made Millions Trading Virtual Loot* and *My Tiny Life: Crime and Passion in a Virtual World*—and has written essays and articles on hackers; computer viruses; online communities; encryption technologies; music pirates; and the heady cultural, political, and philosophical questions that tie these and other digital-age phenomena together. He lives in Chicago.

Dana Goodyear is a staff writer at *The New Yorker* and the author of *Honey and Junk*, a collection of poems. She lives in Los Angeles.

Dan Hill has been working at the forefront of innovative information and communications technology (ICT) since the early 1990s and has been responsible for many popular and critically acclaimed products and services. He has conducted significant strategic work as one of the key architects of a BBC redesigned for the on-demand media age, was Director of Web & Broadcast for *Monocle* magazine, co-organized the "Postopolis" architecture and urbanism conferences in New York and Los Angeles, and runs City of Sound, generally acclaimed as one of the leading architecture and urbanism Web sites.

As a senior consultant at Arup, a global multidisciplinary firm, Dan is exploring the possibilities of ICT from a creative, design-led perspective, rethinking how information changes streets and cities, neighborhoods and organizations, mobility and work, play and public space.

Kevin Kelly is Senior Maverick at *Wired.* He is the author of *Out of Control* and the upcoming book *What Technology Wants.*

Elizabeth Kolbert is a staff writer for *The New Yorker* who reports frequently on science and the environment. She is a recipient of a National Magazine Award, the American Association for the Advancement of Science's magazine-writing award, and the National Academies' communication award. She is also the author of two books, *Field Notes from a Catastrophe: Man, Nature, and Climate Change* and *The Prophet of Love and Other Tales of Power and Deceit.*

Farhad Manjoo is the technology columnist at *Slate* and the author of *True Enough: Learning to Live in Post-Fact Society.* He has also written for *Wired, Playboy*, and the *New York Times.* He lives in San Francisco.

Robin McKie is science editor of *The Observer,* Britain's oldest Sunday newspaper. Apart from covering major science and technology stories for the paper since 1982, he is the author of several books including *Dawn of Man: The Story of Human Evolution* and *African Exodus: The Origins of Modern Humanity* (with Chris Stringer). He is married and lives in London with two children.

Luke O'Brien is a freelance journalist who most recently worked as the Washington, D.C., correspondent for Wired News. He was formerly a video game columnist for Village Voice Media. His work has appeared in *The Atlantic, Rolling Stone,* the *Washington Post Magazine, GOOD,* and *Boston Magazine,* among other publications.

Clay Shirky writes, teaches, and consults on the social effects of the Internet, and especially those places where our social and technological networks overlap. He is on the faculty of NYU's Interactive Telecommunications Program and has consulted for Nokia, Proctor and Gamble, NewsCorp, the BBC, the United States Navy, and Lego. Over the years, his writings have appeared in the *New York Times,* the *Wall Street Journal, Wired,* the *Harvard Business Review,* and *IEEE Computer Review.* His first book, *Here Comes Everybody: The Power of Organizing Without Organizations,* was published by the Penguin Press in 2008.

Adam Sternbergh is an Editor-at-Large for *New York* magazine, where he's worked since 2004. His writing has also appeared in the *New York Times Magazine, GQ,* and on the radio program *This American Life.* His 2006 article "Up with Grups" was included in the anthology *New York Stories: Landmark Writing from Four Decades of* New York Magazine.

Andrew Sullivan is an author, academic, and journalist. He holds a Ph.D. from Harvard in political science and is a former editor of *The New Republic*. His many books include *Virtually Normal: An Argument About Homosexuality* and *The Conservative Soul*. In the summer of 2000, Sullivan became one of the first mainstream journalists to experiment with blogging and soon developed a large online readership with andrewsullivan.com's Daily Dish. Andrew has blogged independently and for Time.com, but in February 2007 Andrew moved his blog to The Atlantic Online, where he now writes daily. The Dish won the Best Blog Award from 2008's Weblog Awards.

David Talbot, chief correspondent at *Technology Review* magazine and former Knight Science Journalism Fellow at MIT, is the recent winner of a *Folio:* Eddie Award for his feature writing, and a past winner of the Overseas Press Club of America Award for international environmental reporting in any medium. He writes and blogs at www.technologyreview.com.

Clive Thompson is a contributing writer for the *New York Times Magazine,* a columnist for *Wired* magazine, and writes for *Fast Company* and *Wired* magazine's Web site, among other places. He blogs about technology, science, and culture at www.collision detection.net.

Sharon Weinberger is a freelance writer specializing in national security and technology. Her writing has appeared in publications such as *Wired,* the *Washington Post Magazine, Nature,* and *Discover.* She was a Knight Science Journalism Fellow at MIT during 2008/2009.

Acknowledgments

Grateful acknowledgment is made to the following authors, publishers, and journals for permission to reprint previously published materials.

"Reflections on Lori Drew, Bullying, and Strategies for Helping Kids" by danah boyd. First published as "Reflections on Lori Drew, Bullying, and Solutions to Helping Kids" in Zephoria.org, September 30, 2008. Reprinted by permission of the author.

"Is Google Making Us Stupid?" by Nicholas Carr. First published in *The Atlantic,* July 2008. Reprinted by permission of the author.

"Rich Man's Burden" by Dalton Conley. First published in the *New York Times,* September 2, 2008. Reprinted by permission of the author.

"Secret Geek A-Team Hacks Back, Defends Worldwide Web" by Joshua Davis. First published in *Wired,* November 24, 2008. Reprinted by permission of the author.

"Mutilated Furries, Flying Phalluses: Put the Blame on Griefers, the Sociopaths of the Virtual World" by Julian Dibbell. First published in *Wired,* January 18, 2008. Copyright © by Julian Dibbell. Reprinted by permission of the author.

"I ♥ Novels" by Dana Goodyear. First published in *The New Yorker,* December 22, 2008. Reprinted by permission of the author.

"The Street as Platform" by Dan Hill. First published in cityofsound .com, February 2008. Reprinted by permission of the author.

"Becoming Screen Literate" by Kevin Kelly. First published in *New York Times Magazine.* Reprinted by permission of the author.

"Dymaxion Man" by Elizabeth Kolbert. First published as "Dymaxion Man: The Visions of Buckminster Fuller" in *The New Yorker,* June 9, 2008. Reprinted by permission of the author.

"The Death of Planned Obsolescence" by Farhad Manjoo. First published in *Slate,* August 11, 2008. Reprinted by permission of the author and *Slate.*

"Isle of Plenty" by Robin McKie. First published in *The Observer* (UK), September 21, 2008. Copyright © 2008 by Robin McKie. Reprinted by permission of the author.

"Spore's Intelligent Designer" by Luke O'Brien. First published in *Slate,* September 11, 2008. Copyright © 2008 by Luke O'Brien. Reprinted by permission of the author.

"Area Eccentric Reads Entire Book." First published in *The Onion,* January 19, 2008. Copyright © 2008 by Onion, Inc. Reprinted by permission of the Onion, Inc., www.theonion.com.

"Gin, Television, and Cognitive Surplus" by Clay Shirky. First published in herecomeseverybody.org. Reprinted by permission of the author.

"The Spreadsheet Psychic" by Adam Sternbergh. First printed in *New York* magazine, October 12, 2008. Reprinted by permission of the author.

"Why I Blog" by Andrew Sullivan. First published in *The Atlantic,* November 2008. Copyright © 2008 by Andrew Sullivan. Reprinted by permission of the author.

"How Obama *Really* Did It" by David Talbot. First published in *Technology Review,* September/October 2008. Reprinted by permission of *Technology Review.*

"Brave New World of Digital Intimacy" by Clive Thompson. First published as "I Am So Digitally Close to You" in *New York Times Magazine,* September 5, 2008. Reprinted by permission of the author.

"Can You Spot the Chinese Nuclear Sub?" by Sharon Weinberger. First published in *Discover Magazine.* Reprinted by permission of the author.

Lightning Source UK Ltd.
Milton Keynes UK
UKHW041352191020
371843UK00002B/255